Integral and Functional Differential Equations

PURE AND APPLIED MATHEMATICS

A Program of Monographs, Textbooks, and Lecture Notes

Contributions to *Lecture Notes in Pure and Applied Mathematics* are reproduced by direct photography of the author's typewritten manuscript. Potential authors are advised to submit preliminary manuscripts for review purposes. After acceptance, the author is responsible for preparing the final manuscript in camera-ready form, suitable for direct reproduction. Marcel Dekker, Inc. will furnish instructions to authors and special typing paper. Sample pages are reviewed and returned with our suggestions to assure quality control and the most attractive rendering of your manuscript. The publisher will also be happy to supervise and assist in all stages of the preparation of your camera-ready manuscript.

LECTURE NOTES
IN PURE AND APPLIED MATHEMATICS

1. *N. Jacobson,* Exceptional Lie Algebras
2. *L.-Å. Lindahl and F. Poulsen,* Thin Sets in Harmonic Analysis
3. *I. Satake,* Classification Theory of Semi-Simple Algebraic Groups
4. *F. Hirzebruch, W. D. Newmann, and S. S. Koh,* Differentiable Manifolds and Quadratic Forms
5. *I. Chavel,* Riemannian Symmetric Spaces of Rank One
6. *R. B. Burckel,* Characterization of C(X) Among Its Subalgebras
7. *B. R. McDonald, A. R. Magid, and K. C. Smith,* Ring Theory: Proceedings of the Oklahoma Conference
8. *Y.-T. Siu,* Techniques of Extension of Analytic Objects
9. *S. R. Caradus, W. E. Pfaffenberger, and B. Yood,* Calkin Algebras and Algebras of Operators on Banach Spaces
10. *E. O. Roxin, P.-T. Liu, and R. L. Sternberg,* Differential Games and Control Theory
11. *M. Orzech and C. Small,* The Brauer Group of Commutative Rings
12. *S. Thomeier,* Topology and Its Applications
13. *J. M. López and K. A. Ross,* Sidon Sets
14. *W. W. Comfort and S. Negrepontis,* Continuous Pseudometrics
15. *K. McKennon and J. M. Robertson,* Locally Convex Spaces
16. *M. Carmeli and S. Malin,* Representations of the Rotation and Lorentz Groups: An Introduction
17. *G. B. Seligman,* Rational Methods in Lie Algebras
18. *D. G. de Figueiredo,* Functional Analysis: Proceedings of the Brazilian Mathematical Society Symposium
19. *L. Cesari, R. Kannan, and J. D. Schuur,* Nonlinear Functional Analysis and Differential Equations: Proceedings of the Michigan State University Conference
20. *J. J. Schäffer,* Geometry of Spheres in Normed Spaces
21. *K. Yano and M. Kon,* Anti-Invariant Submanifolds
22. *W. V. Vasconcelos,* The Rings of Dimension Two
23. *R. E. Chandler,* Hausdorff Compactifications
24. *S. P. Franklin and B. V. S. Thomas,* Topology: Proceedings of the Memphis State University Conference
25. *S. K. Jain,* Ring Theory: Proceedings of the Ohio University Conference
26. *B. R. McDonald and R. A. Morris,* Ring Theory II: Proceedings of the Second Oklahoma Conference
27. *R. B. Mura and A. Rhemtulla,* Orderable Groups
28. *J. R. Graef,* Stability of Dynamical Systems: Theory and Applications
29. *H.-C. Wang,* Homogeneous Banach Algebras
30. *E. O. Roxin, P.-T. Liu, and R. L. Sternberg,* Differential Games and Control Theory II
31. *R. D. Porter,* Introduction to Fibre Bundles
32. *M. Altman,* Contractors and Contractor Directions Theory and Applications
33. *J. S. Golan,* Decomposition and Dimension in Module Categories
34. *G. Fairweather,* Finite Element Galerkin Methods for Differential Equations
35. *J. D. Sally,* Numbers of Generators of Ideals in Local Rings
36. *S. S. Miller,* Complex Analysis: Proceedings of the S.U.N.Y. Brockport Conference
37. *R. Gordon,* Representation Theory of Algebras: Proceedings of the Philadelphia Conference
38. *M. Goto and F. D. Grosshans,* Semisimple Lie Algebras
39. *A. I. Arruda, N. C. A. da Costa, and R. Chuaqui,* Mathematical Logic: Proceedings of the First Brazilian Conference
40. *F. Van Oystaeyen,* Ring Theory: Proceedings of the 1977 Antwerp Conference

41. *F. Van Oystaeyen and A. Verschoren,* Reflectors and Localization: Application to Sheaf Theory
42. *M. Satyanarayana,* Positively Ordered Semigroups
43. *D. L. Russell,* Mathematics of Finite-Dimensional Control Systems
44. *P.-T. Liu and E. Roxin,* Differential Games and Control Theory III: Proceedings of the Third Kingston Conference, Part A
45. *A. Geramita and J. Seberry,* Orthogonal Designs: Quadratic Forms and Hadamard Matrices
46. *J. Cigler, V. Losert, and P. Michor,* Banach Modules and Functors on Categories of Banach Spaces
47. *P.-T. Liu and J. G. Sutinen,* Control Theory in Mathematical Economics: Proceedings of the Third Kingston Conference, Part B
48. *C. Byrnes,* Partial Differential Equations and Geometry
49. *G. Klambauer,* Problems and Propositions in Analysis
50. *J. Knopfmacher,* Analytic Arithmetic of Algebraic Function Fields
51. *F. Van Oystaeyen,* Ring Theory: Proceedings of the 1978 Antwerp Conference
52. *B. Kedem,* Binary Time Series
53. *J. Barros-Neto and R. A. Artino,* Hypoelliptic Boundary-Value Problems
54. *R. L. Sternberg, A. J. Kalinowski, and J. S. Papadakis,* Nonlinear Partial Differential Equations in Engineering and Applied Science
55. *B. R. McDonald,* Ring Theory and Algebra III: Proceedings of The Third Oklahoma Conference
56. *J. S. Golan,* Structure Sheaves over a Noncommutative Ring
57. *T. V. Narayana, J. G. Williams, and R. M. Mathsen,* Combinatorics, Representation Theory and Statistical Methods in Groups: YOUNG DAY Proceedings
58. *T. A. Burton,* Modeling and Differential Equations in Biology
59. *K. H. Kim and F. W. Roush,* Introduction to Mathematical Consensus Theory
60. *J. Banas and K. Goebel,* Measures of Noncompactness in Banach Spaces
61. *O. A. Nielsen,* Direct Integral Theory
62. *J. E. Smith, G. O. Kenny, and R. N. Ball,* Ordered Groups: Proceedings of the Boise State Conference
63. *J. Cronin,* Mathematics of Cell Electrophysiology
64. *J. W. Brewer,* Power Series Over Commutative Rings
65. *P. K. Kamthan and M. Gupta,* Sequence Spaces and Series
66. *T. G. McLaughlin,* Regressive Sets and the Theory of Isols (in press)
67. *T. L. Herdman, S. M. Rankin, III, and H. W. Stech,* Integral and Functional Differential Equations
68. *R. Draper,* Commutative Algebra: Analytic Methods (in press)
69. *W. G. McKay and J. Patera,* Tables of Dimensions, Indices, and Branching Rules for Representations of Simple Lie Algebras

Other Volumes in Preparation

Integral and Functional Differential Equations

edited by

Terry L. Herdman and Harlan W. Stech

Department of Mathematics
Virginia Polytechnic Institute and
State University
Blacksburg, Virginia

and

Samuel M. Rankin, III

Department of Mathematics
West Virginia University
Morgantown, West Virginia

MARCEL DEKKER, INC. New York and Basel

Library of Congress Cataloging in Publication Data
Main entry under title:

Integral and functional differential equations.

(Pure and applied mathematics; 67)
"The Conference on Integral and Functional Differential Equations was held June 18-20, 1979, at West Virginia University in Morgantown, West Virginia"--Pref.
1. Integral equations--Congresses. 2. Functional differential equations--Congresses. I. Stech, Harlan W. [date]. II. Herdman, Terry L. [date]. III. Rankin, Samuel M. [date]. IV. Conference on Integral and Functional Differential Equations (1979 : West Virginia University)
QA431.I47 515.4'5 81-1862
AACR2

MARCEL DEKKER, INC.

270 Madison Avenue, New York, New York 10016

Current printing (last digit):

10 9 8 7 6 5 4 3 2 1

PRINTED IN THE UNITED STATES OF AMERICA

PREFACE

The Conference on Integral and Functional Differential Equations was held June 18-20, 1979, at West Virginia University in Morgantown, West Virginia. The modern facilities of the Evansdale campus provided the setting for a meeting of forty-three participants from twenty-eight colleges and universities, as well as the Oak Ridge National Laboratory. The conference was directed by Samuel M. Rankin, III, and received financial support from West Virginia University and the West Virginia University Foundation. This volume consists of contributions of many of the participants.

The conference proper was preceeded by a few days of informal presentations by some of the participants. In particular, Professor John Nohel (Director of the Mathematics Research Center, University of Wisconsin-Madison) provided a series of four lectures in which he surveyed the development of and recent progress in the area of abstract Volterra integro-differential equations. Part one of this volume is based on these lectures as well as an invited address given during the conference. Other invited lectures were presented by Professors J. K. Hale (Brown University), R. C. MacCamy (Carnegie-Mellon University), V. J. Mizel (Carnegie-Mellon University), R. K. Miller (Iowa State University), and G. F. Webb (Vanderbilt University). These, along with the fourteen contributed papers, document the variety of problems and techniques considered at the conference.

The editors are happy to acknowledge the assistance and encouragement of many individuals and organizations. These include Professor Nohel for his generous contribution to this volume; Dr. Ray Koppelman, Vice President for Energy Studies, Graduate Programs and Research for West Virginia University, without whose help and encouragement this conference would not have taken place; Dr. William G. Collins, Dean, College of Arts and Sciences, West Virginia University, for his support and encouragement; West Virginia University and the West Virginia University Foundation for their organizational and financial support. Also deserving recognition are the

Virginia Tech Department of Mathematics and its Head, C. Wayne Patty, for their assistance in the production of this proceedings. Finally, we are indebted to Elizabeth Kessler for the care and enthusiasm she put into her preparation of the final manuscript.

Terry L. Herdman
Samuel M. Rankin, III
Harlan W. Stech

CONTENTS

CONTRIBUTORS

JOHN A. BURNS, Department of Mathematics, Virginia Polytechnic Institute and State University, Blacksburg, Virginia.

HARVEY DIAMOND, Department of Mathematics, West Virginia University, Morgantown, West Virginia.

RICHARD H. ELDERKIN, Department of Mathematics, Pomona College, Claremont, California.

WILLIAM E. FITZGIBBON*, Department of Mathematics, University of Houston, Houston, Texas.

MELVIN L. HEARD, Department of Mathematics, University of Illinois at Chicago Circle, Chicago, Illinois.

TERRY L. HERDMAN, Department of Mathematics, Virginia Polytechnic Institute and State University, Blacksburg, Virginia.

G. SAMUEL JORDAN, Department of Mathematics, University of Tennessee at Knoxville, Knoxville, Tennessee.

JAMES H. LIGHTBOURNE, III†, Mathematics Department, Pan American University, Edinburg, Texas.

**Current affiliation*: Department of Mathematics, University of California, San Diego, California.

†*Current affiliation*: Department of Mathematics, West Virginia University, Morgantown, West Virginia.

R. C. MacCAMY, Department of Mathematics, Carnegie-Mellon University, Pittsburgh, Pennsylvania.

MOSHE MARCUS*, Department of Mathematics, Carnegie-Mellon University, Pittsburgh, Pennsylvania.

ANTHONY N. MICHEL, Department of Electrical Engineering, Iowa State University, Ames, Iowa.

RICHARD K. MILLER, Mathematics Department, Iowa State University, Ames, Iowa.

VICTOR J. MIZEL, Department of Mathematics, Carnegie-Mellon University, Pittsburgh, Pennsylvania.

JOHN W. NEUBERGER, Department of Mathematics, North Texas State University, Denton, Texas.

JOHN A. NOHEL, Mathematics Research Center, University of Wisconsin-Madison, Madison, Wisconsin.

MARY E. PARROTT†, Department of Mathematical Sciences, Memphis State University, Memphis, Tennessee.

WILLIAM PETERSON, Department of Mathematics, West Virginia University, Morgantown, West Virginia.

WILFRED M. POST, III, Environmental Sciences Division, Oak Ridge National Laboratory, Oak Ridge, Tennessee.

WILLIAM SQUIRE, Department of Aerospace Engineering, West Virginia University, Morgantown, West Virginia.

HARLAN W. STECH, Department of Mathematics, Virginia Polytechnic Institute and State University, Blacksburg, Virginia.

CURTIS C. TRAVIS, Health and Safety Research Division, Oak Ridge National Laboratory, Oak Ridge, Tennessee.

FRANK VERHOFF, Department of Chemical Engineering, West Virginia University, Morgantown, West Virginia.

G. F. WEBB, Mathematics Department, Vanderbilt University, Nashville, Tennessee.

ROBERT L. WHEELER††, Department of Mathematics, University of Missouri, Columbia, Missouri.

Current affiliation: Department of Mathematics, Technion-Israel Institute of Technology, Haifa, Israel.

†*Current affiliation*: Department of Mathematics, University of South Florida, Tampa, Florida.

††*Current affiliation*: Department of Mathematics, Virginia Polytechnic Institute and State University, Blacksburg, Virginia.

CONFERENCE PARTICIPANTS

J. Michael Berman
University of Rhode Island/
Virginia State College

Craig Beasley
North Texas State University

John A. Burns
Virginia Tech

P. Clément
University of Technology, Delft,
The Netherlands

Richard Datko
Georgetown University

Harvey Diamond
West Virginia University

Rodney D. Driver
University of Rhode Island

Richard H. Elderkin
Pamona College

William E. Fitzgibbon
University of Houston

Jack K. Hale
Brown University

Kenneth Hannsgen
Virginia Tech

Melvin Heard
University of Illinois at Chicago
Circle

Terry L. Herdman
Virginia Tech

G. Samuel Jordan
University of Tennessee

Thomas R. Kiffe
Texas A & M University

R. W. Leggett
Oak Ridge National Laboratory

Marshall J. Leitman
Case Western Reserve University

James H. Lightbourne, III
Pan American University

R. C. MacCamy
Carnegie-Mellon University

Joseph M. McDonough
West Virginia University

Moshe Marcus
Carnegie-Mellon University

Richard K. Miller
Iowa State University

Victor J. Mizel
Carnegie-Mellon University

John Neuberger
North Texas State University

John A. Nohel
University of Wisconsin-Madison

Mary E. Parrott
Memphis State University

Stefano L. Paveri-Fontana
SUNY at Buffalo

William Peterson
West Virginia University

Wilfred M. Post
Oak Ridge National Laboratory

Samuel M. Rankin, III
West Virginia University

James A. Reneke
Clemson University

Stephen Saperstone
George Mason University

Juan Jorge Schaffer
Carnegie-Mellon University

Henry Simpson
University of Tennessee

William Squire
West Virginia University

Harlan W. Stech
Virginia Tech

Michael Stecher
Texas A & M University

Ken Swick
Queens College

Curtis Travis
Oak Ridge National Laboratory

Glenn F. Webb
Vanderbilt University

Robert L. Wheeler
University of Missouri

Lynn Roy Williams
Indiana University at South Bend

Donald F. Young
Agnes Scott College

Integral and Functional Differential Equations

Part I: LECTURE SERIES

NONLINEAR VOLTERRA EQUATIONS FOR HEAT FLOW IN MATERIALS WITH MEMORY

John A. Nohel

Mathematics Research Center
University of Wisconsin-Madison
Madison, Wisconsin

0. OVERVIEW

These notes on lectures delivered at West Virginia University preceding the conference on Volterra and Functional Differential Equations in June 1979, concern the existence, uniqueness, positivity, boundedness, and asymptotic behavior as $t \to \infty$ of solutions of the nonlinear Volterra equation

$$u(t) + (b*Au)(t) \ni f(t) \qquad (0 \le t < \infty) \tag{V}$$

The general setting for V is as follows: $b:[0,\infty) \to R$ is a given kernel, A is a m-accretive, possibly multivalued, operator on a real Banach space X, $f:[0,\infty) \to X$ is a given function, and * denotes the convolution; the integral in V is taken in the sense of Bochner. The special cases of A maximal monotone on a real Hilbert space H, and $A = \partial\phi$, the subdifferential of a proper, convex, lower semicontinuous function $\phi:H \to (-\infty,\infty]$ will also play a prominent role, primarily in the boundedness and asymptotic theory for (V).

It should be observed that if $b \equiv 1$ and $f \in W^{1,1}(0,\infty;X)$, where W denotes the usual Sobolev space, equation (V) is formally equivalent to the evolution problem

$$\frac{du}{dt} + Au \ni f' \qquad (0 < t < \infty),\ u(0) = u_0 = f(0) \tag{E}$$

Thus the theory for (V) is to a considerable extent a generalization of the theory of evolution equations, and uses most of the techniques for the latter combined with techniques for Volterra equations developed in recent years.

Section 1 is primarily intended for motivation. Beginning from simple physical principles and following ideas of B. D. Coleman, M. E. Gurtin, R. C. MacCamy, W. Noll, J. W. Nunziato, and A. C. Pipkin, we derive equation (V) as one possible mathematical model for nonlinear heat flow in a homogeneous body of material with memory. While the derivation is restricted to one space dimension, the modification for heat flow in a homogeneous body Ω of isotropic material in R^2 or R^3 is also indicated. One purpose of Section 1 is to arrive at a physically reasonable set of conditions concerning the kernel b in (V), under which one can expect boundedness and the kind of asymptotic behavior of solutions studied later in Section 3. The kernel b in (V) does not arise directly from physical principles; rather, it is the case that b is expressed in terms of two physically measurable quantities (at least in principle) about which one can make appropriate assumptions based on physical considerations. The types of assumptions made concerning the operator A and the function f are also motivated. In particular, A is a nonlinear second-order elliptic partial differential operator in the space variables which incorporates the boundary conditions, and f depends on the external heat supply, the initial temperature distribution in the body and the history of temperature in the body. If $b \equiv 1$, then in the application (V) is equivalent to the classical nonlinear heat equation in an ordinary body.

Another purpose of Section 1 is to point out that heat flow in certain materials with memory can also be modelled by integrodifferential equations other than (V). Limitations of space do not permit us to discuss the qualitative aspects of these alternative models in any detail; however, references to relevant existing literature are given.

Section 2 discusses the theory of existence, uniqueness, and continuous dependence of global solutions of (V), both in the general case of a real Banach space and in the special case of a real Hilbert space and for $A = \partial\phi$ where the results are stronger. The development is primarily based on recent joint work with M. G. Crandall [26], and partly on a recent paper by G. Gripenberg [34]. References to earlier and related literature are given.

Section 3 develops the theory of boundedness and asymptotic behavior of solutions of (V) as $t \to \infty$, under assumptions partly motivated by the heat flow problem formulated in Section 1; application of the theory to this problem is given. The development is based on forthcoming joint work with P. Clément and R. C. MacCamy [20]. References to other pertinent literature are given.

Section 4, based on recent and forthcoming joint work with P. Clément, [18,19], as well as recent work by Clément [17], discusses the existence, boundedness and asymptotic behavior as $t \to \infty$ of *positive* solutions of (V) under assumptions which are also motivated by the heat flow problem in Section 1. This problem is then used to illustrate the theory. The reader should recall that it is classical that solutions of the heat equation are positive, if the initial temperature distribution and the external heat supply are positive. Some of the results of this section were presented at the West Virginia Conference by Clément.

1. VOLTERRA EQUATIONS OCCURRING IN HEAT FLOW IN MATERIALS WITH MEMORY

1.1. Introductory Remarks

The purpose of Section 1 is to derive from physical considerations several mathematical models for nonlinear heat flow in materials with memory. For simplicity we shall limit most of our considerations to heat flow in one space dimension and remark about the situation in the multidimensional case. The primary objective is to arrive at equation (V) in Section 1.2 below under physically reasonable assumptions on the kernel b, the operator A and the forcing term f. This particular mathematical model motivates many of the considerations in Sections 2, 3, and 4. In Sections 1.3 and 1.4 we shall also derive two other models for heat flow in materials with memory which have been and are being studied, but due to limitations of space we will only refer the reader to the relevant mathematical literature for their analysis.

The mathematical models are derived from the following general considerations of energy balance for heat transfer in body B in $R^n (n = 1,2,3)$. If $\varepsilon(t,x)$ represents the internal energy, $\overline{q}(t,x)$ represents the heat flux, and $h(t,x)$ represents the heat supply, where t is the time and x is the position in the body, then the energy balance equation is

$$\varepsilon_t = -\mathrm{div}\bar{q} + h \qquad (t > 0,\ x \in B) \tag{1.1}$$

The classical linear heat equation which accurately describes heat transfer by conduction in many materials is derived from (1.1) by assuming that the heat flux obeys Fourier's law:

$$\bar{q} = -c_0 \text{ grad } u$$

where $c_0 > 0$ is the constant thermal conductivity and u represents the temperature in the body at time t and position x. It is also assumed that the internal energy depends linearly on the temperature

$$\varepsilon = \varepsilon_0 + b_0 u$$

where $b_0 > 0$ is the constant heat capacity and $\varepsilon_0 \geq 0$ is a constant. The energy balance (1.1) then yields the linear heat equation

$$b_0 u_t = c_0 \nabla^2 u + h$$

which adequately describes the evolution of temperature in most homogeneous and isotropic rigid bodies.

However, some materials exhibit memory effects (materials of fading memory type, see Coleman and Mizel [23]) for which the classical theory is unable to account. Heat flow in such material is modelled by assuming that the internal energy ε and the heat flux q are respectively functionals (rather than functions) of the temperature and of the gradient of temperature. The considerations which follow are based on extensive research by Coleman, Gurtin, Noll, Pipkin, MacCamy, Mizel, and Nunziato (see especially Coleman [21], Coleman and Gurtin [22], Coleman and Mizel [23], Gurtin and Pipkin [39], MacCamy [57], [58], [59], Nunziato [68]).

1.2. A Model for Nonlinear Heat Flow in a Material with Memory

We consider nonlinear heat flow in a homogeneous bar of unit length of material with memory with the temperature $u = u(t,x)$ maintained at zero at $x = 0$ and $x = 1$. We shall assume that the history of u is prescribed for $t \leq 0$ and $0 \leq x \leq 1$. The equation satisfied by u in such a material is derived from the assumptions that the internal energy ε and the heat flux q are functionals (rather than functions) of u and of the gradient of u respectively. According to the theory developed by Coleman, Gurtin, Noll, Pipkin, MacCamy and Nunziato (see e.g. MacCamy [57], [59] and Nunziato [68]) for heat flow in materials of fading memory type the functionals ε and q are taken respectively as:

$$\varepsilon(t,x) = b_0 u(t,x) + \int_0^t \beta(t - s)u(s,x)ds \qquad (t \geq 0,\ 0 < x < 1) \tag{2.1}$$

$$q(t,x) = -c_0\sigma(u_x(t,x)) + \int_0^t \gamma(t - s)\sigma(u_x(s,x))ds \tag{2.2}$$

$(t \geq 0,\ 0 < x < 1)$

In writing the functions ε and q we have assumed for simplicity and without loss of generality that the history of the temperature u is prescribed as zero for $t < 0$ (if this is not the case and if the history of u is sufficiently smooth for $t < 0$ and $0 < x < 1$, this has the effect of altering the forcing term h in equation (2.3) below and consequently also G in (2.5) below--by additional known forcing terms). In (2.1), (2.2) $b_0 > 0$, $c_0 > 0$ are given constants, $\beta,\gamma : [0,\infty) \to R$ are given sufficiently smooth functions which we call the internal energy and heat flux relaxation functions respectively.

The real function $\sigma : R \to R$ in (2.2) will be assumed to satisfy the assumptions (σ) of Lemma 2.3 below. It should be noted that the case $\sigma(r) \equiv r$ gives rise to the linear model derived in Nunziato [68], and that (2.2) is one reasonable generalization of the heat flux functional for nonlinear heat flow.

In the physical literature it is customary to define

$$\alpha(t) = b_0 + \int_0^t \beta(s)ds \qquad (0 \leq t < \infty)$$

as the internal energy relaxation function; thus $\alpha(0) = b_0$ and $\alpha'(t) = \beta(t)$. Similarly,

$$\kappa(t) = c_0 - \int_0^t \gamma(s)ds \qquad (0 \leq t < \infty)$$

is defined as the heat flux relaxation function, so that $\kappa(0) = c_0$ and $\kappa'(t) = -\gamma(t)$. Then (2.1), (2.2) are replaced respectively by

$$\varepsilon(t,x) = \varepsilon_0 + \alpha(0)u(t,x) + \int_0^t \alpha'(t - s)u(s,x)ds, \qquad \varepsilon_0 \geq 0 \tag{2.1'}$$

$$q(t,x) = -\kappa(0)\sigma(u_x) - \int_0^t \kappa'(t - s)\sigma(u_x(s,x))ds \tag{2.2'}$$

see [68, (5.13), (5.11)] where the linear case $\sigma(r) \equiv r$ is considered. The quantity $\alpha(0)$ is called the instantaneous heat capacity while $\alpha(\infty)$ is the equilibrium heat capacity with similar names for the $\kappa(0)$ and $\kappa(\infty)$. It is shown in [68] that $\alpha(\infty) \geq \alpha(0) > 0$, and that $\kappa(0)$, $\kappa(\infty)$ are nonnegative.

In the physical literature β, γ are usually assumed to be decaying exponentials with positive coefficients. As we shall see the theory developed in Sections 2-4 permits a much greater generality, and we shall merely have to require that $\beta(0) > 0$, $\gamma(0) > 0$, that β and $\gamma \in L^1(0,\infty)$, and that

$$b_0 + \int_0^t \beta(\tau)d\tau > 0, \qquad c_0 - \int_0^t \gamma(\tau)d\tau > 0 \qquad (0 \le t < \infty)$$

which corresponds to the physically reasonable assumptions $\alpha(t) > 0$, $\kappa(t) > 0$, $0 \le t < \infty$. We shall also assume that the conditions

$$b_0 + \mathrm{Re}\,\hat{\beta}(i\eta) > 0 \qquad (\eta \in R) \tag{PW}$$

$$c_0 - \int_0^\infty \gamma(\tau)d\tau > 0 \tag{γ}$$

where $\hat{\beta}(i\eta) = \int_0^\infty \beta(t)\exp(-i\eta t)dt$, are satisfied; assumption (γ) states that $\kappa(\infty) > 0$. The above assumptions will be motivated presently. Remark 4.8 in Section 3 below shows that the physically reasonable assumptions $b_0 + \int_0^t \beta(\tau)d\tau > 0$ and $c_0 - \int_0^t \gamma(\tau)d\tau > 0$ $(0 \le t < \infty)$ are actually not essential for the theory developed in Section 3 to apply.

If $h = h(t,x) \in L^1_{loc}(0,\infty;L^2(0,1))$ represents the external heat supply added to the rod for $t \ge 0$ and $0 < x < 1$, and if $u(0,x) = u_0(x)$, $0 < x < 1$, is the given initial temperature distribution, the law of balance of heat (1.1) shows that in one space dimension the temperature u satisfies the initial-boundary value problem

$$\begin{cases} \frac{\partial}{\partial t}[b_0 u + \beta * u] = c_0\sigma(u_x)_x - \gamma*\sigma(u_x)_x + h \\ \quad (0 < t < \infty,\ 0 < x < 1) \\ u(0,x) = u_0(x) \qquad (0 < x < 1) \\ u(t,0) = u(t,1) \equiv 0 \qquad (t \ge 0) \end{cases} \tag{2.3}$$

where subscripts denote differentiation with respect to x and where * denotes the convolution on [0,t]. Note that in an ordinary material $\beta = \gamma \equiv 0$, and (2.3) becomes the nonlinear heat equation in one space dimension.

The next task is to transform (2.3) to the equivalent form (V) below which will be used for the analysis. Define

$$C(t) = c_0 - \int_0^t \gamma(\tau)d\tau \qquad (0 \le t < \infty) \tag{2.4}$$

$$G(t,x) = b_0 u_0(x) + \int_0^t h(\tau,x)d\tau \qquad (0 \le t < \infty,\ 0 < x < 1) \tag{2.5}$$

Noting that

$$\frac{\partial}{\partial t}(C*\sigma(u_x)_x)(t,x) = c_0\sigma(u_x(t,x))_x - (\gamma*\sigma(u_x)_x)(t,x)$$

and integrating (2.3) using the initial condition, and (2.5), one obtains the equivalent Volterra equation (to (2.3)):

$$b_0u(t,x) + (\beta*u)(t,x) = (C*\sigma(u_x)_x)(t,x) + G(t,x)$$

$$(0 < t < \infty,\ 0 < x < 1) \tag{2.6}$$

where u satisfies the boundary conditions $u(t,0) = u(t,1) \equiv 0$ $(t \geq 0)$.

We next define the nonlinear operator A formally by the relation

$$Au = -\sigma(u_x)_x \qquad \text{where} \qquad u(t,0) = u(t,1) \equiv 0 \tag{A}$$

In order not to interrupt this development we postpone a precise definition of A to Lemma 2.3 below. Then the Volterra equation (2.6) has the abstract form

$$b_0u + \beta*u + C*Au = G \qquad (0 \leq t < \infty) \tag{V_1}$$

To transform (V_1) to the equivalent form (V) below, define $\rho : [0,\infty) \to R$ (called the resolvent kernel of β) to be the unique solution of the linear Volterra equation

$$b_0\rho(t) + (\beta*\rho)(t) = -\frac{\beta(t)}{b_0} \quad (0 \leq t < \infty) \tag{ρ}$$

It is standard that if $b_0 > 0$ and $\beta \in L^1_{loc}[0,\infty)$, equation ($\rho$) has a unique solution $\rho \in L^1_{loc}(0,\infty)$. Applying the variation of constants formula for Volterra equations [63]

$$(b_0y + \beta*y = g \iff y = \frac{g}{b_0} + \rho*g)$$

finally yields that (V_1) is equivalent to the abstract equation

$$u + b*Au = f \qquad (0 \leq t < \infty) \tag{V}$$

with the definitions

$$b(t) = \frac{C(t)}{b_0} + (\rho*C)(t) \qquad (0 \leq t < \infty) \tag{b}$$

$$f(t) = \frac{G(t,\cdot)}{b_0} + (\rho*G)(t,\cdot) \qquad (0 \leq t < \infty) \tag{f}$$

Similar considerations show that for heat flow in a bounded homogeneous body Ω of isotropic material with memory in R^2 or R^3 with a smooth boundary Γ, the temperature u will also satisfy the abstract Volterra equation (V) with the kernel b and forcing term f given exactly as above, but with the nonlinear operator A defined precisely in Remark 2.4.

We next comment on the significance of the assumptions concerning β, γ as well as (PW) and (γ). Since the relaxation functions β and γ are generally taken as decaying exponentials with positive coefficients in the physical literature, it is certainly reasonable to assume that $\beta,\gamma \in L^1(0,\infty)$ and that $\beta(0) > 0$, $\gamma(0) > 0$. We next motivate the assumption that $b_0 + \int_0^t \beta(\tau)d\tau > 0$ $(0 \leq t < \infty)$. A similar reasoning motivates $c_0 - \int_0^t \gamma(\tau)d\tau > 0$ $(0 \leq t < \infty)$. Consider the internal energy ε defined by (2.1) and suppose that the temperature u is maintained at zero up to time t_0 and at a state $\overline{u}_1(x) > 0$ $(0 < x < 1)$ for $t > t_0$. One would then expect the internal energy to be positive for $t > t_0$. If the function β is positive for $t \geq 0$ this is automatically the case. However if not, the assumption $b_0 + \int_0^t \beta(\tau)d\tau > 0$ $(0 \leq t < \infty)$ is natural in view of the fact that in this situation

$$\varepsilon(t,x) = \overline{u_1}(x)\left(b_0 + \int_{t_0}^t \beta(\tau)d\tau\right) \qquad (t_0 \leq t < \infty)$$

Since $\beta \in L^1(0,\infty)$ equation (2.1) shows that ε is bounded whenever u is bounded. The assumption (PW) implies that $b_0 + \int_0^\infty \beta(t)dt > 0$ (take $\eta = 0$); thus if $u(x,t)$ tends to an equilibrium state $\overline{u}(x) > 0$ as $t \to \infty$, (2.1) implies that the corresponding limiting internal energy $\overline{\varepsilon}(x) > 0$ as is to be expected. For physical reasons it is also to be expected that if ε is bounded the temperature should be bounded. Applying the variations of constants formula to (2.1) yields

$$u(t,x) = \frac{\varepsilon(t,x)}{b_0} + (\rho * \varepsilon)(t,x) \qquad (0 \leq t < \infty,\ 0 < x < 1)$$

where ρ is the resolvent kernel of β defined by equation (ρ). Thus to have u bounded whenever ε is bounded it is sufficient to require that $\rho \in L^1(0,\infty)$. But by the Paley-Wiener theorem [69] applied to equation (ρ), $\beta \in L^1(0,\infty)$ implies that $\rho \in L^1(0,\infty)$ if and only if

$$b_0 + \hat{\beta}(z) \neq 0 \qquad \text{for} \qquad \operatorname{Re} z \geq 0$$

The condition (PW) now results from taking the real part of this expression, noting that for physical reasons one wants $b_0 + \hat{\beta}(0) > 0$, and arguing as in the proof of Lemma 2.2, Section 3.

To motivate assumption (γ) suppose that $u(t,x) \to \overline{u}(x)$ as $t \to \infty$ and that $d/dx\, \overline{u}(x) > 0$, implying that $\sigma(d\overline{u}/dx) > 0$ (see assumptions (σ) below). One then expects that the limiting heat flux $\overline{q}(x)$ in equation (2.2) is strictly negative, if the process being modelled represents "forward" heat flow; condition (γ) insures that this is the case.

We shall next see that the physically reasonable assumptions $b_0 > 0$, $c_0 > 0$, (PW) and (γ), together with some mild additional technical assumptions, imply that the kernel b in the Volterra equation (V) defined in (b) satisfies the assumptions

$$b(t) = b_\infty + B(t), \qquad b(0) > 0, \qquad b_\infty > 0$$

$$B, B' \in L^1(0,\infty) \tag{H_b}$$

these will play an important role in the application of the boundedness and asymptotic theory developed in Section 3.3 and in the application of that theory to heat flow described by the problem (2.3) above in Section 3.4. One has the following result whose elementary proof is omitted:

LEMMA 2.1. Let $b_0 > 0$, $c_0 > 0$, β, γ, $t\beta$, $t\gamma \in L^1(0,\infty)$, and let assumptions (PW) and (γ) be satisfied. Define

$$b_\infty = \frac{c_0 - \int_0^\infty \gamma(t)dt}{b_0 + \int_0^\infty \beta(t)dt} \tag{2.7}$$

$$B(t) = \frac{C(t)}{b_0} + (\rho * C)(t) - b_\infty \tag{2.8}$$

where $C(t) = c_0 - \int_0^t \gamma(\tau)d\tau$, and ρ is the resolvent of β uniquely defined by equation (ρ). Then $b_\infty > 0$ and $B, B' \in L^1(0,\infty)$, and $b(t) = b_\infty + B(t)$ satisfies (H_b) with $b(0) = c_0/b_0 > 0$, $B(0) = c_0/b_0 - b_\infty > 0$.

The next elementary result gives physically reasonable sufficient conditions on the relaxation function β, the initial temperature distribution u_0, and the external heat supply h in order that the forcing term f in (V) defined by equation (f) will satisfy the assumption

$$f(t) = f_\infty + F(t), \qquad F \in W^{1,2}_{loc}([0,\infty);H), \qquad F' \in L^2(0,\infty;H) \tag{H_f}$$

where H is the real Hilbert space $L^2(0,1)$ and W is the usual Sobolev space. Assumption (H_f) will play an important role in the boundedness and asymptotic theory of Section 3.

LEMMA 2.2. Let $H = L^2(0,1)$ and $u_0 \in H_0^1(0,1)$. Let $\beta \in L^1(0,\infty) \cap L^2(0,\infty)$ and let assumption (PW) be satisfied. Finally, assume that

$$h \in L^1(0,\infty;H) \cap L^2(0,\infty;H) \tag{h}$$

Then the function $f : [0,\infty) \times (0,1) \to H$, defined by equations (f), (G), where ρ is the resolvent of β, satisfies $f \in W^{1,2}_{loc}(0,\infty;H)$ and $f(0,x) = u_0(x) \in H_0^1(0,1)$. Moreover,

$$f(t,x) = f_\infty(x) + F(t,x) \qquad (0 \le t < \infty,\ 0 < x < 1)$$

where

$$f_\infty(x) = (b_0u_0(x) + \textstyle\int_0^\infty h(\tau,x)d\tau)(\frac{1}{b_0} + \int_0^\infty \rho(\tau)d\tau) \tag{2.9}$$

$$\begin{aligned} F(t,x) &= \frac{G(t,x)}{b_0} + (\rho*G)(t,x) - f_\infty(x) = -\frac{1}{b_0}\int_t^\infty h(\tau,x)d\tau \\ &\quad - \int_0^t \rho(t-s)\int_s^\infty h(\tau,x)d\tau ds - \int_t^\infty \rho(\tau)d\tau(b_0u_0(x) + \int_0^\infty h(\tau,x)d\tau) \end{aligned} \tag{2.10}$$

and $\partial F/\partial t \in L^2(0,\infty;H)$. If in addition $t\beta \in L^1(0,\infty)$ and $th \in L^1(0,\infty;H)$, then $F \in L^2(0,\infty;H)$.

Sketch of Proof of Lemma 2.2. The assumptions $\beta \in L^1(0,\infty)$ and (PW), together with the Paley-Wiener theorem [69], applied to the resolvent equation (ρ) imply that $\rho \in L^1(0,\infty)$. But then the assumption $\beta \in L^2(0,\infty)$ and the fact that $\rho \in L^1(0,\infty)$ imply that also $\rho \in L^2(0,\infty)$ from the resolvent equation. These facts combined with the definition of f in (f) and assumption (h) yield the formulae (2.9) for f_∞ and (2.10) for F given in the statement, as well as $f \in W^{1,2}_{loc}(0,\infty;H)$. From formula (2.10) one easily proves that

$$\begin{aligned} &\frac{\partial F}{\partial t}(t,x) = \frac{1}{b_0} h(t,x) + b_0u_0(x)\rho(t) + (\rho*h)(t,x) \\ &(0 \le t < \infty,\ 0 < x < 1) \end{aligned} \tag{2.11}$$

then $\partial F/\partial t \in L^2(0,\infty;H)$ follows from $h \in L^2(0,\infty;H)$ and $\rho \in L^1(0,\infty) \cap L^2(0,\infty)$. Finally, (PW) and $t\beta \in L^1(0,\infty)$, together with $\rho \in L^1(0,\infty)$ imply that $t\rho \in L^1(0,\infty)$ from the resolvent equation. This, together with the assumption $th \in L^1(0,\infty)$ and routine estimates applied to the formula (2.10) yield $F \in L^2(0,\infty;H)$. This completes the proof.

The next task is to give a precise definition of the operator A in the abstract equation (V) for the heat flow problem under study. Let $H = L^2(0,1)$ be the real Hilbert space of square integrable functions on (0,1). Let $\sigma : R \to R$ satisfy the assumptions

$$\sigma \in C^1(R), \qquad \sigma(0) = 0, \qquad \sigma'(\xi) \ge p_0 > 0 \qquad (\xi \in R) \tag{σ}$$

for some $p_0 > 0$. Define $W : R \to R^+$ by

$$W(r) = \int_0^r \sigma(\xi)d\xi \qquad (\ge \frac{p_0}{2} r^2) \qquad (r \in R)$$

and define $\phi : H \to (-\infty,+\infty]$ by

$$\phi(y) = \begin{cases} \int_0^1 W(\frac{dy}{dx}(x))dx & \text{if} \quad y \in H_0^1(0,1) \\ +\infty & \text{otherwise} \end{cases} \tag{2.12}$$

LEMMA 2.3. Let the assumptions (σ) be satisfied and let $\phi : H \to (-\infty,+\infty]$ be the function defined by (2.12). Then ϕ is convex, l.s.c. and proper on H, and

$$Ay = \partial\phi(y) = -\frac{d}{dx}\sigma(\frac{dy}{dx})$$

$$D(\partial\phi) = \{y \in H_0^1(0,1) : \frac{d}{dx}\sigma(\frac{dy}{dx}) \in L^2(0,1)\}$$

where $\partial\phi$ denotes the subdifferential. Moreover, $\phi(y) \geq 0$, $\phi(y) \to \infty$ as $\int_0^1 |y(x)|^2 dx \to \infty$, and $(y,\partial\phi(y)) \geq p_0\pi^2 \int_0^1 |y(x)|^2 dx$, where $(\cdot,\cdot)$ denotes the scalar product in H.

Sketch of Proof of Lemma 2.3. The first result is standard see Brézis ([14],[15]). To prove the last two conclusions let $y \in H_0^1(0,1)$; then from the definition of ϕ and the Poincaré inequality one has

$$\phi(y) \geq \frac{p_0}{2}\int_0^1 |\frac{dy}{dx}(x)|^2 dx \geq \frac{p_0}{2}\pi^2 \int_0^1 |y(x)|^2 dx \geq 0$$

and $\phi(y) \to \infty$ as $\int_0^1 |y(x)|^2 dx \to \infty$. Moreover,

$$(y,\partial\phi(y)) = -\int_0^1 y(x)\frac{d}{dx}\sigma(\frac{dy}{dx}(x))dx$$

and an integration by parts, $y \in H_0^1(0,1)$ and the Poincaré inequality give

$$(y,\partial\phi(y)) = \int_0^1 \frac{dy}{dx}(x)\sigma(\frac{dy}{dx}(x))dx \geq p_0\int_0^1 |\frac{dy}{dx}(x)|^2 dx$$

$$\geq p_0\pi^2 \int_0^1 |y(x)|^2 dx$$

While our considerations of the heat flow problem are primarily in one space dimension, we indicate how to define the nonlinear operator $A = \partial\phi$ in (V) for the heat flow problem in two or three space dimensions so that the function ϕ satisfies the conclusions of Lemma 2.3 under the physically reasonable assumption (λ) below.

REMARK 2.4. Let Ω be a bounded domain in R^n(for heat flow n = 2 or 3) with smooth boundary Γ. Let $\lambda : R^+ \to R$ be a given smooth function satisfying the assumption

$$\begin{cases} \lambda(0) > 0, \quad \text{there exists } p_0 > 0 \text{ such that } \lambda(\xi) \geq p_0 \text{ and} \\ \xi\lambda'(\xi) + \lambda(\xi) \geq p_0 \quad (\xi \in R) \end{cases} \tag{λ}$$

Define $\Lambda : R \to R^+$ by

$$\Lambda(r) = \int_0^r \xi\lambda(\xi)d\xi \qquad (\geq \frac{p_0}{2} r^2) \qquad (r \in R)$$

Let $H = L^2(\Omega)$ and define

$$\phi(u) = \begin{cases} \int_\Omega \Lambda(|\nabla u|)dx & \text{if} \quad u \in H_0^1(\Omega) \\ +\infty & \text{otherwise} \end{cases}$$

Then it is readily verified that $\phi : H \to (-\infty,+\infty]$ is convex, l.s.c., and proper on H and

$$Au = \partial\phi(u) = -\nabla\cdot(\lambda(|\nabla u|)\nabla u)$$

with

$$D(\partial\phi) = \{u \in H_0^1(\Omega) : \nabla\cdot(\lambda(|\nabla u|)\nabla u) \in L^2(\Omega)\}$$

Clearly $\phi(u) \geq 0$ $(u \in H)$ and by the Poincaré inequality $\phi(u) \to \infty$ as $|u| \to \infty$. Using integration by parts and the Poincaré inequality one also has the coercivity condition

$$(Au,u) \geq kp_0|u|_H^2$$

where $k > 0$ is the constant in the Poincaré inequality.

The results of Lemmas 2.1-2.3 and of Remark 2.4 will be used in Section 3.4, to discuss the global existence, uniqueness, boundedness and asymptotic behavior of the solution of the initial-boundary value problem (2.3) using the theory which will be developed for the equivalent abstract Volterra equation (V).

1.3. A General (Parabolic) Heat Flow Model

We consider the same heat flow problem as in Section 1.2. In this model we assume that the internal energy is given by the functional (2.1), but we assume the following more general form of the heat flux functional

$$q(t,x) = -\psi(u_x) - \int_0^t a(t - s)\sigma(u_x(s,x))ds \qquad (t \geq 0) \tag{3.1}$$

where we again assume that the history of temperature is prescribed as zero for $t < 0$ and $0 \leq x \leq 1$. The real function $\psi : R \to R$, $\psi(0) = 0$ satisfies the same assumptions as σ in Lemma 2.3. If $\psi(\cdot) = c_0\sigma(\cdot)$, and $a = -\gamma$ (i.e. in the notation of (2.2') $\kappa'(t) = a(t)$) (3.1) reduces to (2.2). We assume that β, a, and the external heat supply h satisfy the same type of smoothness assumptions respectively as β and γ in Section 1.2. We assume again that

$$b_0 + \int_0^t \beta(\tau)d\tau > 0 \qquad (0 \leq t < \infty)$$

and that $a(0) > 0$. For physical reasons it is also reasonable to assume that

$$\psi(\xi) + \left(\int_0^t a(\tau)d\tau\right)\sigma(\xi) \begin{cases} > 0 & \text{if} \quad \xi > 0, \quad t \geq 0 \\ < 0 & \text{if} \quad \xi < 0, \quad t \geq 0 \end{cases}$$

and that

$$\psi(\xi) + \left(\int_0^\infty a(\tau)d\tau\right)\sigma(\xi) \begin{cases} > 0 & \text{if} \quad \xi > 0 \\ < 0 & \text{if} \quad \xi < 0 \end{cases}$$

Applying the energy balance equation (1.1), and using (2.1), (3.1) and h shows that the temperature u satisfies the initial-boundary value problem

$$\begin{cases} b_0 u_t + \beta(0)u + \beta'*u = \psi(u_x)_x + a*\sigma(u_x)_x + h \\ \qquad (0 < t < \infty, \ 0 < x < 1) \\ u(t,0) = u(t,1) \equiv 0 \qquad (t \geq 0) \\ u(0,x) = u_0(x) \qquad (0 < x < 1) \end{cases} \tag{3.2}$$

Note that if $a = \beta \equiv 0$ (3.2) reduces to the nonlinear heat equation. Defining the operators A, B, L by the relations:

$$Au = -\sigma(u_x)_x \qquad \text{where} \qquad u(t,0) = u(t,1) \equiv 0$$

$$Bu = -\psi(u_x)_x \qquad \text{where} \qquad u(t,0) = u(t,1) \equiv 0$$

$$Lu = \beta(0)u + \beta'*u$$

and taking (without loss of generality) $b_0 = 1$, one sees that the problem (3.2) has the abstract form

$$\begin{cases} \dfrac{du}{dt} + Bu + a*Au + Lu = h \qquad (0 < t < \infty) \\ u(0) = u_0 \end{cases} \tag{V_L'}$$

If the functions σ and ψ satisfy the assumptions (σ), Lemma 2.3 shows that A and B are subdifferentials of proper, convex, l.s.c. functions defined on the real Hilbert space $H = L^2(0,1)$. If the relaxation function β has $\beta' \in L^1(0,\infty)$, the linear operator L is well defined.

The abstract problem (V_L') has been investigated by a number of authors combining techniques of Volterra equations and the theory of monotone operators. With $\beta \equiv 0$ V. Barbu [5], [7], Barbu and Malik [11] studied the problem of global existence; a more complete existence theory for considerably more general kernels a, as well as a discussion of boundedness and asymptotic behavior, was developed by Crandall, Londen, and Nohel [28]. The latter, see [28, p. 717], also permits in the existence theory Lipschitz type perturbations of the operator A; this essentially covers the case of the specific operator L in the present application; the details of this generalization of the existence theory have recently been worked out by M. J. Luo as a part of a forthcoming Ph.D. thesis at the University of Wisconsin-Madison.

The general assumptions for the existence theory in [28] concerning the operators A and B are roughly speaking as follows. The operators A and B are subdifferentials of proper, convex, l.s.c. functions defined on a real Hilbert space H; B dominates A in a certain precise sense (see [28], inequality (1.7); the case $B = kA$, $k > 0$ is not excluded), and B satisfies a compactness assumption (see (1.8) in [28]); this compactness condition excludes the possibility $B \equiv 0$. The kernel a satisfies an abstract condition (see [28; (1.10)]) which is shown to hold for two physically important classes of kernels:

$$\begin{cases} a(0) > 0, \quad a \text{ locally absolutely continuous on } [0,\infty) \\ a' \text{ locally of bounded variation on } [0,\infty) \end{cases} \tag{a_1}$$

$$\begin{cases} a(0) > 0, \quad a \in C[0,\infty) \cap C^2(0,\infty), \quad \text{and} \\ a \text{ nonnegative, nonincreasing, convex on } [0,\infty) \end{cases} \tag{a_2}$$

It should be noted that no satisfactory uniqueness theorem has to date been discovered for the general problem (V_L') with $B \neq kA$, $k > 0$ a constant, even when $L \equiv 0$.

Other approaches to the study of (V_L') include Aizicovici [1,2,3] and Gripenberg [32], [36]. While these studies are in a more general setting than those referred to above, they do not seem to shed new light on the physical problem (3.2), and indeed the existence results of Aizicovici do not include those of [28]. An extension of the latter to nonconvolution kernels has been studied by Rennolet [71]. A semigroup approach to a special case of (V_L') has been investigated by Vrabie [83]. Another interesting "parabolic" heat flow problem involving an analysis of evolution inequalities has recently been studied by J. Naumann [64].

1.4. A Hyperbolic Nonlinear Volterra Equation for Heat Conduction with Finite Wave Speeds

The parabolic models for heat flow in materials with memory formulated in Sections 1.2 and 1.3 both predict that a thermal disturbance at any point of the body is instantly felt everywhere in the body (through not with equal strength). This implies that finite discontinuities propagate with infinite speed. This situation is unrealistic for some materials, particularly at low temperatures.

Gurtin and Pipkin [39], see also Nunziato [68], have proposed a model for heat flow which exhibits a finite speed of propagation; the study of the linear model $\sigma(r) \equiv r$. We present briefly such a nonlinear model which has been investigated by MacCamy [59] by the method of characteristics, and by Dafermos and Nohel [29] by an energy method. Another interesting variant of the energy method was recently developed by Staffans [80].

Consider the heat flow problem of Section 1.2. Define the internal energy by the functional (2.1), with β satisfying the conditions of Section 1.2. In place of (2.2) assume that the heat flux is given by the functional

$$q(t,x) = -\int_0^t c(t-s)\sigma(u_x(s,x))ds \qquad (0 < t < \infty,\ 0 < x < 1) \tag{4.1}$$

where the relaxation function $c \in L^1(0,\infty)$, $c(0) > 0$, $\int_0^t c(s)ds \geq 0$ $(0 \leq t < \infty)$, $\int_0^\infty c(s)ds > 0$, and where the real function σ satisfies assumptions (σ) of Lemma 2.3 (actually σ, smooth, $\sigma(0) = 0$, $\sigma'(0) > 0$ is sufficient for the development in [29]). Note that from (4.1) the heat flux depends only on the history of the gradient of u, and is independent of the present value of the gradient of u. Evidently, this model of the heat flux results from (2.2) by taking $c_0 = 0$ and by replacing γ by $-c$, or from (2.2') by taking $\kappa(0) = 0$, $\kappa'(t) = c(t)$. The model (4.1) for the heat flux also results by taking $\psi \equiv 0$ in (3.1), a case which is excluded in the theory and the referenced accompanying mathematical literature in Section 1.3.

Applying the energy balance (1.1) and using (2.1), (4.1) and the external heat supply h leads to the equation

$$b_0u_t + \frac{\partial}{\partial t}\beta * u = c*\sigma(u_x)_x + h \qquad (0 < t < \infty,\ 0 < x < 1)$$

Noting that $\beta * u = u * \beta$, carrying out the differentiation, and imposing the boundary and initial conditions leads to the following initial-boundary value problem for the heat flow problem under the assumptions of this section

$$\begin{cases} b_0 u_t + \beta * u_t = c * \sigma(u_x)_x + h(t,x) - \beta(t) u_0(x) \\ \quad (0 < t < \infty,\ 0 < x < 1) \\ u(t,0) = u(t,1) \equiv 0 \qquad (t \geq 0) \\ u(0,x) = u_0(x) \end{cases} \tag{4.2}$$

The problem (4.2) is transformed to the more standard abstract form (V_H') below as follows. As in Section 1.2 define the resolvent kernel ρ of β by the linear Volterra equation (ρ); by assumption (PW) $\rho \in L^1(0,\infty)$. Define the function $a : [0,\infty) \to R$ by

$$a(t) = \frac{1}{b_0} c(t) + (\rho * c)(t) \qquad (0 \leq t < \infty)$$

define the function $g : [0,\infty) \times (0,1) \to R$ by

$$g(t,x) = \frac{1}{b_0} (h(t,x) - \beta(t) u_0(x)) + \rho * (h(t,x) - \beta(t) u_0(x))$$

$$(0 < t < \infty,\ 0 < x < 1)$$

Finally define the operator A as in Section 1.2. Applying the variation of constants formula for Volterra equations, and these definitions to (4.2) shows that (4.2) is equivalent to the abstract nonlinear Volterra equation

$$\begin{cases} \dfrac{du}{dt} + a * Au = g(t,\cdot) \qquad (0 < t < \infty) \\ u(0) = u_0(\cdot) \end{cases} \tag{V_H'}$$

It is easy to see (compare Lemma 2.1) that under the present assumptions on β and c, $a \in L^1(0,\infty)$ and that

$$\int_0^\infty a(t)dt = \frac{\int_0^\infty c(t)dt}{b_0 + \int_0^\infty \beta(t)dt} > 0$$

if also $\beta(0) > 0$, $c(0) > 0$, $c'(0) \leq 0$, then $a(0) > 0$ and $a'(0) < 0$. It is also evident that the forcing term $g \in L^2(0,\infty;L^2(0,1))$ if $h \in L^2(0,\infty;L^2(0,1))$ and also $\beta \in L^2(0,\infty)$.

Notice that if $\beta \equiv 0$ and $\gamma \equiv \gamma(0) > 0$ (or equivalently $a \equiv \gamma(0)/b_0$) the problem (4.2) reduces to the nonlinear wave equation problem

$$\begin{cases} b_0 u_{tt} = \gamma(0)\sigma(u_x)_x + h_t & (0 < t < \infty,\ 0 < x < 1) \\ u(t,0) = u(t,1) \equiv 0 & (t \geq 0) \\ u(0,x) = u_0(x), \quad u_x(0,x) = u_1(x) = g(0,x) \end{cases} \tag{W}$$

If the real function σ is "genuinely nonlinear" ($\sigma''(\xi) \not\equiv 0$, $\xi \in R$), Lax [50] has shown that (W) fails to have global smooth (C^2) solutions in time, even if $h \equiv 0$, no matter how smooth and "small" one takes the initial functions u_0, u_1. If the function σ is convex the derivatives of the solution u of (W) develop singularities due to the crossing of characteristics in finite time ("shocks").

The objective of the analysis by MacCamy [59] (which uses Riemann invariants and is therefore restricted to one space dimension), and a different analysis by Dafermos and Nohel [29, Theorem 4.1], Staffans [80], which are applicable to several space dimensions and both of which use energy methods, is to show that under the present assumptions on β and γ (together with some technical ones and some other physically reasonable ones which imply that the kernel a is strongly positive on $[0,\infty)$), the integral in (V_H') has the effect of a frictional damping mechanism which prevents the formation of shocks, provided the data u_0 and the forcing term g are sufficiently smooth and small in certain H norms. This analysis leads to global existence, uniqueness, and decay of smooth solutions of the problem (4.2) for sufficiently smooth and small data u_0 and h (see especially [29, Theorems 4.1, 6.1]; for a physically reasonable two-dimensional heat flow problem with the same kernel a see [29, Theorem 7.1]). It is also evident from the analysis in [29, Section 3] dealing with the local existence and uniqueness of the problem (4.2), resp. (V_H'), that solutions of (4.2), resp. (V_H'), possess the property of finite speed of propagation. For an analysis of the existence and uniqueness of classical solutions for "small" data in the simpler case of a nonlinear wave equation with frictional damping see Nishida [65], and Nohel [66].

We remark also that the abstract problem (V_H') has recently been studied by S. -O. Londen [55], [56] for a class of kernels a which are positive, decreasing, and convex on $(0,\infty)$ and which satisfy the crucial (for his method) condition $a'(0+) = -\infty$. His method is a significant generalization of that of Crandall, Londen, and Nohel [28] for the parabolic problems discussed briefly in Section 1.3. However, the assumption

$a'(0+) = -\infty$ is inappropriate in the present physical context; moreover, the type of solution obtained by Londen in [55] and [56] need not be regular in the sense of smooth solutions, and no decay results of solutions comparable to [29] are obtainable by his methods. It should also be remarked that for the linear problem (V_H') (e.g. $Au = -\nabla^2 u$), interesting and useful results using deep results of Shea and Wainger [73] have been obtained in a series of papers by K. B. Hannsgen [41-47] and by Carr and Hannsgen [16].

Finally, we note the model problem (V_H') is similar to a particular model problem for nonlinear viscoelastic motion in one space dimension in which, however, the kernel a has the form $a(t) = a_\infty + A(t)$, $a_\infty > 0$, A positive, decreasing, convex on $[0,\infty)$, and for which the analysis is considerably more complicated. This problem has been extensively studied by MacCamy [60], Dafermos and Nohel [29], and Staffans [80]; see also an interesting asymptotic result by Staffans [81] motivated by this problem.

2. EXISTENCE AND UNIQUENESS OF SOLUTIONS OF ABSTRACT VOLTERRA EQUATIONS

2.1. Introduction

In Section 2 we study the abstract nonlinear Volterra equation

$$u(t) + b*Au(t) \ni f(t) \qquad (0 \leq t \leq T) \tag{V}$$

where $T > 0$ is arbitrary, in the setting: A is an m-accretive (possibly multivalued) operator in a real Banach space X, the given kernel b is a real absolutely continuous function on $[0,T]$, $b*g(t) = \int_0^t b(t-s)g(s)ds$ with the integral in (V) interpreted as the usual Bochner integral, and the given function $f \in W^{1,1}(0,T;X)$ where $W^{1,1}$ is the usual Sobolev space.

We treat the problem of existence, uniqueness, dependence on data, and regularity of solutions of (V) on $[0,T]$ by means of a simple method developed jointly with M. G. Crandall [26] to which the reader is referred for more details; the results obtained for (V) generalize and simplify considerably earlier work on existence and uniqueness obtained by Barbu [6], [8], London [54], Gripenberg [30], for the case $X = H$ a real Hilbert space. A different approach to the study of (V) in the same general setting was developed independently by Gripenberg [31,32]. The general theory will be used in Sections 3 and 4. We will also comment on the special cases: (i) A maximal monotone on H, and (ii) $A = \partial\phi$, where $\phi : H \to (-\infty,+\infty]$

is a proper, convex, l.s.c. function and $\partial\phi$ denotes the subdifferential of ϕ (see Brézis [14]); these special cases will be important in Section 3. We shall also consider briefly a recent generalization of [26] by Gripenberg [34], which will also be used in Section 4.

Our method involves reducing the study of (V) to that of an equivalent functional differential equation of the form

$$\frac{du}{dt} + Au \ni G(u) \qquad (0 \le t \le T)$$
$$u(0) = x = f(0) \tag{FDE}$$

where $G : C([0,T]; \overline{D(A)}) \to L^1(0,T;X)$ is a particular mapping, and developing the theory for (FDE). Our results are also directly applicable to certain integrodifferential equations studied by MacCamy [58] via a Galerkin argument which necessitates further restrictions.

We observe that if $b \equiv 1$, equation (V) is equivalent to the evolution problem

$$\frac{du}{dt} + Au \ni f', \qquad u(0) = x = f(0), \qquad (0 \le t \le T) \tag{E}$$

Our method of studying (FDE) consists of generalizing known results for (E) due primarily to Benilan [13]; the latter are reviewed in Section 2.2. We recall also that the initial-boundary value problem for a linear or nonlinear diffusion problem is a special case of (E).

2.2. Preliminaries on Evolution Equations

For further background and details of this section we refer the reader to [7], [25], [26]. Let X be a real Banach space with norm $|\cdot|$. A mapping $A : X \to 2^X$ is called an *operator* in X; its *domain* $D(A) = \{x \in X : Ax \neq \phi\}$ and its *range* $R(A) = \cup\{Ax : x \in D(A)\}$; A is single-valued if Ax is a singleton. An operator A in X is *accretive* iff $J_\lambda = (I + \lambda A)^{-1}$ is a contraction in X for $\lambda \ge 0$. It follows immediately: A is accretive iff

$$\| (x_1 + \lambda y_1) - (x_2 + \lambda y_2) \| \ge \| x_1 - x_2 \|$$
$$\text{for} \quad y_i \in Ax_i \quad (i = 1,2) \tag{2.1}$$

An operator A in X is called *m-accretive* iff A is accretive and $R(I + \lambda A) = X$ for $\lambda > 0$.

We shall be concerned with applying some known facts about the abstract evolution equation

$$\frac{dv}{dt} + Av \ni g, \qquad v(0) = x \tag{E_g}$$

to the study of (FDE). We assume throughout that $g \in L^1(0,T;X)$, $T > 0$.

DEFINITION 2.1. A function $v : [0,T] \to X$ is a strong solution of (E_g) on $[0,T]$ if $v(0) = x$, $v \in C([0,T];X) \cap W^{1,1}(0,T;X)$, $v(t) \in D(A)$ a.e. on $[0,T]$ and there exists $w \in Av$ such that $w(t) \in Av(t)$ and $v'(t) + w(t) = g(t)$ a.e. on $[0,T]$.

DEFINITION 2.2. $v : [0,T] \to X$ is a weak solution of (E_g) on $[0,T]$ if there is a sequence $\{(v_n,g_n)\}_{n=1}^{\infty} \in C([0,T];X) \times L^1(0,T;X)$ such that v_{n_1} is a strong solution of (E_{g_n}) on $[0,T]$ and $(v_n,g_n) \to (v,g)$ in $C([0,T];X) \times L^1(0,T;X)$.

For our considerations we require a third concept of solution of (E_g), namely the notion of integral solution. First, let $[\ ,\ ]_\lambda : X \times X \to R$ be defined for $\lambda > 0$ by

$$[x,y]_\lambda = \frac{1}{\lambda}(\| x + \lambda y \| - \| x \|)$$

which is a nondecreasing function of λ. Define

$$[x,y]_+ = \lim_{\lambda \downarrow 0} [x,y]_\lambda = \inf_{\lambda > 0} [x,y]_\lambda$$

$$[x,y]_- = \lim_{\lambda \uparrow 0} [x,y]_\lambda = \sup_{\lambda < 0} [x,y]_\lambda$$

Thus $\| x + \lambda y \| \geq \| x \|$ for $\lambda \geq 0$ iff $[x,y]_+ \geq 0$, so that A is accretive iff

$$[x_1 - x_2, y_1 - y_2]_+ \geq 0 \qquad \text{for} \qquad y_i \in Ax_i \tag{2.2}$$

DEFINITION 2.3. $v : [0,T] \to X$ is an integral solution of (E_g) on $[0,T]$ if $v \in C([0,T];X)$ and

$$\| v(t) - x \| - \| v(s) - x \| \leq \int_s^t [v(\alpha) - x,\ g(\alpha) - y]_+ \, d\alpha \tag{2.3}$$

for $t > s$, $(t,s \in [0,T])$, $x \in D(A)$ and $y \in Ax$. We note since $|[x,y]_+| \leq \| y \|$, and since $g \in L^1(0,T;X)$, the integral in (2.3) is well defined. A straightforward calculation, see [25], shows that the notion of integral solution only makes sense when A is accretive. We shall apply the following result on existence, uniqueness, dependence on data, and regularity about integral solutions of (E_g) due to Benilan [13].

THEOREM A. If A is m-accretive, $x \in \overline{D(A)}$, and $g \in L^1(0,T;X)$ then (E_g) has a unique integral solution $v \in C([0,T]; \overline{D(A)})$ on $[0,T]$, and if $v,\hat{v}$ are integral solutions of (E_g), $(E_{\hat{g}})$ on [0,T] corresponding to initial values $x,\hat{x}$ respectively then

$$\|\hat{v}(t) - v(t)\| \leq \|x - \hat{x}\| + \int_0^t \|\hat{g}(\sigma) - g(\sigma)\|d\sigma,$$

$$0 \leq t \leq T \tag{2.4}$$

Moreover, if $g \in BV([0,T];X)$ and $x \in D(A)$, then

$$\|v(\xi) - v(\eta)\| \leq |\xi \quad \eta|\{\|g(0^+) - y\| + var(g : [0,t])\} \tag{2.5}$$

for $y \in Ax$, and $0 \leq \xi, \eta \leq t$, $t \in [0,T]$. In particular, the integral solution v is Lipschitz continuous. If, in addition, X is reflexive, then v is a strong solution of (E_g) on [0,T].

2.3. Discussion of Existence and Uniqueness Results

We shall reduce the study of existence and uniqueness of solutions of the nonlinear Volterra equation (V) on [0,T] to studying the abstract functional differential equation

$$\frac{du}{dt} + Au \ni G(u) \qquad (0 \leq t \leq T)$$
$$u(0) = x \tag{FDE}$$

where A is a given m-accretive operator on X, and where G is a given mapping

$$G : C([0,T]; \overline{D(A)}) \to L^1(0,T;X)$$

Let $v = H(g)$ denote the unique integral solution of (E_g). A solution of (FDE) is by definition a function $u \in C([0,T]; \overline{D(A)})$ such that $u = H(G(u))$. By analogy with Definition 2.1, we say that u is a strong solution of (FDE) on [0,T] if $u(0) = x$, $u \in W^{1,1}(0,T;X) \cap C([0,T]; \overline{D(A)})$ and if $u'(t) + Au(t) \ni G(u)(t)$ a.e. on [0,T].

Let $b \in L^1(0,T;R)$, $F \in L^1(0,T;X)$. We shall say that u is a strong solution of the Volterra equation (V) on [0,T] if $u \in L^1(0,T;X)$ and if there exists $w \in L^1(0,T;X)$ such that $w(t) \in Au(t)$ and $u(t) + b*w(t) = F(t)$ a.e. on [0,T]. One establishes the following equivalence between strong solutions of (FDE) with a particular G and strong solutions of (V):

PROPOSITION 3.1. Let $b \in AC([0,T];R)$, $b' \in BV([0,T];R)$, $F \in W^{1,1}(0,T;X)$ and $b(0) = 1$. Let u be a strong solution of (V) on [0,T]. Then u is a strong solution of (FDE) on [0,T] with the identifications:

$$\begin{cases} \text{(i)} \quad G(u)(t) = f'(t) - r*f'(t) - a(0)u(t) \\ \qquad\quad - r(t)x + \int_0^t u(t-s)dr(s) \\ \text{(ii)} \quad x = f(0) \\ \text{(iii)} \quad a = b' \\ \text{(iv)} \quad r \in L^1(0,T;R) \text{ is defined by } r + a*r = a \end{cases} \tag{3.1}$$

Conversely, let $r \in BV([0,T];R)$, $f' \in L^1(0,T;X)$, $x \in \overline{D(A)}$ and G be given by (3.1) (i). Let u be a strong solution of (FDE) on [0,T]. Then u is a strong solution of (V) on [0,T], where

$$\begin{cases} \text{(i)} \quad f(t) = x + \int_0^t f'(s)ds \\ \text{(ii)} \quad b(t) = 1 + \int_0^t a(s)ds \\ \text{(iii)} \quad a - a*r = r \end{cases} \tag{3.2}$$

We remark that if $b(t) \equiv 1$ and $F \in W^{1,1}(0,T;X)$, then the Volterra equation (V) is equivalent to the evolution equation (E_g) where $g = f'$ and where the initial value $x = f(0)$.

The proof of Proposition 1 is straightforward. The assumptions on b and F permit differentiation a.e. on [0,T] of a strong solution u of (V). The differentiated equation is then "solved" for Au by means of the resolvent kernel r associated with $a = b'$, see (3.1)(iv), and the variation of constants formula for Volterra equations [63]. A known result [12] yields that $a \in BV([0,T];R)$ implies that $r \in BV([0,T];R)$, a fact which is used in arriving at the formula (3.1)(i) for G(u). The converse is proved by reversing the steps. A part of Proposition 3.1 which motivates our approach is contained in MacCamy [58] who, however, then studied (FDE) by an entirely different approach.

We remark that here we have chosen to define the resolvent kernel by (3.1)(iv), rather than by $r + a*r = -a$ as was done in [26]. This is more convenient for the theory in Chapter 4, and only causes a change of signs in the formula (3.1)(i) of some of the terms in G(u)(t). Recall that if r is defined by (3.1)(iv), then the solution of the linear Volterra equation $w + a*w = v$ is given by $w = v - r*v$, while with the alternate definition of r, w would be given by $w = v + r*v$, as was used in [26].

We next use Benilan's theorem about solutions of (E_g) to obtain some general results concerning existence, uniqueness, dependence on data, and regularity of solutions of (FDE) of independent interest and use them to deduce corresponding results about solutions of (V).

THEOREM 3.2. Assume that A is m-accretive, $x \in \overline{D(A)}$, and let $G : C([0,T]; \overline{D(A)}) \to L^1(0,T;X)$ satisfy

$$\left\{\begin{array}{l} \| G(u) - G(v) \|_{L^1(0,t;X)} \le \int_0^t \gamma(s) \| u - v \|_{L^\infty(0,s;X)} ds \\ \text{for some } \gamma \in L^1(0,T;R^+), \quad 0 \le t \le T, \quad \text{and} \\ \quad u,v \in C([0,T];\overline{D(A)}) \end{array}\right. \tag{3.3}$$

Then (FDE) has a unique solution $u \in C([0,T]; \overline{D(A)})$ on $[0,T]$.

We remark that assumption (3.3) implies that the value of G(u) at $t \in [0,T]$ depends only on the restriction of u to $[0,t]$. The idea of the proof is very simple. Let $v = H(g)$ denote the unique integral solution of (E_g) on $[0,T]$, $g \in L^1(0,T;X)$. We seek a fixed point of the map $K : C([0,T];D(A)) \to C([0,T]; \overline{D(A)})$ defined by $K(u) = H(G(u))$. By property (2.4) of integral solutions

$$\| K(u)(t) - K(v)(t) \| \le \int_0^t \| G(u)(s) - G(v)(s) \| ds$$

$$(0 \le t \le T)$$

for $u,v \in C([0,T]; \overline{D(A)})$, $u(0) = v(0) = x$. With assumption (3.3) it is now an easy matter to show that K^j is a strict contraction on $C([0,T]; \overline{D(A)})$ for j sufficiently large, so that the map K has a unique fixed point. For details see [26].

Under further assumptions one can apply the second part of Benilan's theorem to obtain greater regularity of solutions of (FDE).

THEOREM 3.3. In addition to the assumption of Proposition 3.1 assume that there is a function $k : [0,\infty) \to [0,\infty)$ such that

$$\left\{\begin{array}{l} \operatorname{var}(G(u) : [0,t]) \le k(R)(1 + \operatorname{var}(u : [0,t])) \\ \text{and } \| G(u)(0^+) \| \le k(R) \qquad (0 \le t \le T) \end{array}\right. \tag{3.4}$$

whenever $u \in C([0,T]; \overline{D(A)})$ is of bounded variation and $\| u \|_{L^\infty(0,T;X)} \le R$. If $x \in D(A)$, then the solution u of (FDE) is Lipschitz continuous on $[0,T]$. If X is also reflexive, then the solution u is a strong solution of (FDE) on $[0,T]$.

For the proof of Theorem 3.3 one defines $u_0 : [0,T] \to X$ by $u_0(t) = x$ and $u_{n+1} = K(u_n) = H(G(u_n))$, $n = 0,1,\ldots$. These iterates converge uniformly and are uniformly bounded on $[0,T]$. By Benilan's theorem and assumption (3.4) one shows that there exists a constant $c > 0$ such that

$$\mathrm{var}(u_{n+1} : [0,t]) < c(1 + \int_0^t \mathrm{var}(u_n : [0,s])ds)$$

for $0 \le t \le T$, so that $\mathrm{var}(u_{n+1} : [0,t]) \le c \exp(ct)$. Thus $\{\mathrm{var}(u_n : [0,T])\}$ and by (3.4) $\{\mathrm{var}(G(u_n)) : [0,T]\}$ are both bounded, and $\{u_n\}$, and hence also $u = \text{unif} \lim u_n$, is Lipschitz continuous on $[0,T]$. For more details see [26].

Finally, the solution u of (FDE) depends on the data A, G, x in the following sense:

THEOREM 3.4. Let the assumptions of Proposition 3.1 be satisfied. Let m-accretive operators A_n in X, mappings $G_n : C([0,T];X) \to L^1(0,T;X)$, and $x_n \in \overline{D(A)}$ be given for $n = 1,2,\ldots$. Assume that the inequality (3.3) holds for G replaced by G_n, with the same γ, for $n = 1,2,\ldots$, and $u,v \in C([0,T];\overline{D(A)})$. For $u \in C([0,T];\overline{D(A)})$ assume that $\lim_{n\to\infty} G_n(u) = G(u)$ in $L^1(0,T;X)$, $\lim x_n = x \in \overline{D(A)}$, and

$$\lim_{n\to\infty} (I + \lambda A_n)^{-1}z = (I + \lambda A)^{-1}z \qquad (z \in X,\ \lambda > 0) \tag{3.5}$$

Let $u_n \in C([0,T];\overline{D(A_n)})$ be solutions of (FDE) on $[0,T]$ with A replaced by A_n, G replaced by G_n, x replaced by x_n, and let $u \in C([0,T];\overline{D(A)})$ be the solution of (FDE) on $[0,T]$. Then $\lim_{n\to\infty} u_n = u$ in $C([0,T];X)$.

The proof of Theorem 3.4 follows from the observation that under our assumptions the mapping $K(A,x,G)(u) = H(A,x,G(u))$ of Theorem 3.2 has the property that in the iterate K^j, which is a strict contraction for some j, both j and the contraction constant depend only on the function γ of (3.3), and the latter is assumed to be uniform in n; for details see [26].

We shall next apply Theorems 3.2, 3.3, 3.4 to study the nonlinear Volterra equation (V). If b and F in (V) satisfy the assumptions of Proposition 3.1, it follows from the definition of G in (3.1)(i) that

$$\| G(u)(t) - G(v)(t) \| \le (|r(0^+)| + \mathrm{var}(r : [0,t]))\| u - v \|_{L^\infty(0,t;X)}$$

where r is the resolvent kernel corresponding to $b' = a$ (recall that $a \in BV([0,T];R) \implies r \in BV([0,T];R)$). Thus assumption (3.3) of Theorem 3.2 is satisfied with

$$\gamma(s) = |r(0)| + \text{var}(r : [0,s])$$

Moreover, if $f' \in BV([0,T];X)$, (3.1)(i), (ii), imply

$$\text{var}(G(u) : [0,t]) \leq C(1 + \text{var}(u : [0,t])) \qquad (0 \leq t \leq T)$$

and $\| G(u)(0^+) \| \leq C$, where C is a constant depending on $f(0)$, $f'(0^+)$, $\text{var}(f' : [0,T])$, $r(0^+)$, and $\text{var}(r : [0,T])$; thus assumption (3.4) of Theorem 3.3 is satisfied.

Let $\lambda > 0$ and define the Yosida approximation A_λ of the m-accretive operator A on X by

$$A_\lambda = \frac{1}{\lambda}(I - J_\lambda), \qquad J_\lambda = (I + \lambda A)^{-1}$$

$A_\lambda : X \to X$ is Lipschitz continuous with Lipschitz constant $\frac{1}{\lambda}$, so a simple contraction argument shows that the approximating problem

$$u_\lambda + b*A_\lambda u_\lambda = f \tag{V_λ}$$

has a unique strong solution u_λ on $[0,T]$, under the assumptions: $b \in L^1(0,T;R)$, and $f \in L^1(0,T;X)$. By Proposition 3.1 u_λ is a strong solution of

$$\frac{du_\lambda}{dt} + A_\lambda u_\lambda = G(u_\lambda), \qquad u_\lambda(0) = F(0) \tag{FDE_λ}$$

One also had $\lim\limits_{\lambda \downarrow 0} (I + \mu A_\lambda)^{-1} z = (I + \mu A)^{-1} z$, for $\mu > 0$, $z \in X$.

These considerations lead to the following result about solutions of (V).

THEOREM 3.5. (See [26, Theorem 4). Let $T > 0$ and let the following assumptions be satisfied:

b is absolutely continuous on $[0,T]$, $b(0) > 0$, and $b' \in BV[0,T]$

A is m-accretive on X

$f \in W^{1,1}(0,T;X)$, $f(0) \in \overline{D(A)}$

Then equation (V) has a unique (generalized) solution $u \in C([0,T]; \overline{D(A)})$ in the sense that (i) u is a unique solution of (FDE) on $[0,T]$ with the identifications (3.1) and (ii) $u = \lim\limits_{\lambda \uparrow 0} u_\lambda$ in $C([0,T];X)$, where u_λ are strong solutions of the approximating equation (V_λ) on $[0,T]$.

If, moreover, $f' \in BV(0,T ;X)$ and $f(0) \in D(A)$, then the generalized solution u is Lipschitz continuous on $[0,T]$. If also X is reflexive, then u is a strong solution of (V) on $[0,T]$.

REMARKS 3.6. (i) We remark that if the Volterra equation (V) has a strong solution u on [0,T] under the assumptions of Theorem 3.3, then from Theorem 3.3 and Proposition 3.1, $\lim_{\lambda \downarrow 0} u_\lambda = u$ in $C([0,T];X)$ exists, where u_λ is the strong solution of the approximating equation (V_λ). However, under our assumptions the solutions u_λ of (V_λ) converge to a limit u as $\lambda \downarrow 0$, whether or not (V) has a strong solution. For this reason we refer to the solution u of (V) of Theorem 3.5 as the *generalized solution* of V on [0,T]. Moreover, we note that if the assumption $b' \in BV[0,T]$ only holds on $[0,T_0]$ for some fixed $T_0 > 0$, then by a standard translation argument (see [55]) the solution can be extended to [0,T].

(ii) A precise estimate giving the dependence of the generalized solution u of (V) on the data b and F is established in Theorem 5 of [26].

(iii) The assumption b(0) = 1 in Theorem 3.5 and in Proposition 3.1 is no loss of generality, provided b(0) > 0. For if b(0) > 0, defining $\tilde{b} = (b(0))^{-1}b$ and $\tilde{A} = b(0)A$ one has $b*Au = \tilde{b}*\tilde{A}u$.

(iv) Our method can be used to study the nonconvolution Volterra equation

$$u(t) + \int_0^t b(t,s)Au(s)ds \ni f(t) \qquad (0 \le t \le T)$$

where A and f are as in Theorem 3.5, provided the kernel b, which is defined on the region $\{(t,s) : 0 \le s \le t \le T\}$, is sufficiently smooth and $b(t,t) > 0$. The technique for doing this is outlined in [26], and is carried out in detail by C. Rennolet [72]. For different nonconvolution equation results see Gripenberg [37], [38].

(v) In Theorem 3.5 above and 3.7 below it is important to note that the generalized (or strong) solution $u \in C([0,T];X)$, and therefore $u \in L^p(0,T;X)$ for $p \ge 1$.

(vi) If the assumptions of Theorems 3.5 above and 3.7 below are satisfied for every T > 0, then the conclusions hold on $[0,\infty)$.

(vii) The relation of Theorem 3.5 to other literature is explained in [26].

(viii) An interesting situation not covered by the theory discussed above arises if the kernel b in (V) has the property

$$\lim_{t \to 0^+} \frac{b(t) - b(0)}{t} = +\infty$$

In the case $A = \partial\phi$, where ϕ is convex, l.s.c. and proper, G. Gripenberg [36] extends the theory by replacing the assumption $b' \in BV[0,T]$ in Theorem 3.5 above by:

there exist $T_0 > 0$, $c_0 > 0$ such that if $0 < t < T_0$,

$\mathrm{Var}(b';[t,T_0]) \leq c_0 \log t^{-1}$

Recently G. Gripenberg [34, Theorem 2] obtained the following important generalization of Theorem 3.5 which will be used in Section 4. Such a result was established by Clément and Nohel [18] for the much simpler case $b_1 \equiv 0$ below and A a linear operator.

THEOREM 3.7. Let the assumptions concerning f, A in Theorem 3.5 be satisfied. Let the kernel $b = b_1 + b_2$ in (V), where b_1 satisfies the assumptions of b in Theorem 3.5, and where $b_2 \in L^1(0,T)$ and b_2 is positive, nonincreasing, and $\log b_2$ is convex on $(0,T)$. Then the Volterra equation (V) has a generalized solution $u \in C([0,T];X)$.

Finally, we close Section 2 with two important special cases. The first deals with the case of the operator A being maximal monotone on a real Hilbert space H and is a direct consequence of Theorem 3.5. Recall that if $X = H$, A is m-accretive iff A is maximal monotone.

THEOREM 3.8. Let b satisfy the assumptions of Theorem 3.5 for every $T > 0$. Let A be maximal monotone on H. Let $f \in W^{1,2}_{loc}(0,\infty;H)$, $f(0) \in \overline{D(A)}$. Then (V) has a unique generalized solution $u \in C([0,\infty);\overline{D(A)})$. If, in addition $f' \in BV_{loc}([0,\infty);H)$, $f(0) \in D(A)$, then $u \in W^{1,2}_{loc}(0,\infty;H)$ and u is a strong solution of (V) on $[0,\infty)$.

REMARK 3.9. By another theorem of Gripenberg [34, Theorem 1], this result also applies to (V) with kernels $b = b_1 + b_2$ which satisfy the assumptions of Theorem 3.7.

The second special case of Theorem 3.5 deals with $X = H$ a real Hilbert space and the operator $A = \partial\phi$ (the subdifferential of ϕ), where $\phi : H \to (-\infty,+\infty]$ is a proper, convex, l.s.c. function. The proof of the next result follows by combining Theorem 3.5 with known results for evolution equations (for details see [26, Section 4]).

THEOREM 3.10. Let the kernel b satisfy the assumption of Theorem 3.5 for every $T > 0$. Let $A = \partial\phi$, where $\phi : H \to (-\infty,+\infty]$ is proper, l.s.c., and convex. Let $f \in W^{1,2}_{loc}(0,\infty;H)$. If $f(0) \in \overline{D(\phi)}$, then (V) has a unique strong solution u on $[0,\infty)$ such that $\sqrt{t}\, u' \in L^2_{loc}(0,\infty;H)$; if $f(0) \in D(\phi)$, then $u' \in L^2_{loc}(0,\infty;H)$.

REMARK 3.11. In a different direction Kiffe and Stecher [48] study existence and uniqueness of $L^2(0,T)$ solutions of (V) in a Hilbert space setting. They assume that $f \in L^2(0,T;H)$ and they use techniques of Barbu [5] and Londen [54] to obtain their results without any differentiability assumptions on the forcing term f, but at the expense of drastically restricting the growth of the maximum monotone operator A in (V). In fact, this restriction rules out the important possibility that A is a nonlinear differential operator in the spatial variables, and therefore, their results cannot be applied to the physical problem in Section 1.

An interesting and different variant of (V) was recently studied by Kiffe [49]. He obtains existence of global solutions of the equation

$$u + b*[Au + g(u)] \ni f \qquad (0 \le t \le T)$$

where A can be a nonlinear differential operator in the space variables and where the perturbation g is a discontinuous real function which is not necessarily monotone and satisfies certain growth conditions. The kernel b satisfies assumptions similar to those in this chapter; while $b' \in BV[0,T]$ is not assumed, certain monotonicity is required of b. The forcing term $f \in W^{1,2}$; the operator $A = \partial\phi$, where the function ϕ is as in Theorem 3.10. The function ϕ satisfies a compactness assumption, and $f(0) \in D\phi$.

REMARK 3.12. Other interesting variants of the abstract equation (V), motivated by the heat flow problem formulated in Section 1.2, have been studied by V. Barbu [8], [10], and by H. Attouch and A. Damlamian [4]. In [8,10] Barbu generalizes the dependence of the internal energy ε on the temperature; this leads him to study the equation

$$Bu + b*Au \ni f \qquad (0 \le t < \infty)$$

where B is a strictly monotone operator. In [4] the domain of the operator A in (V) is allowed to depend on time t; for the heat flow problem formulated in Section 1.2, the temperature u is prescribed at each time t outside a body $\Omega(t)$ in x space. Assuming that $\Omega(t)$ depends smoothly on t, the temperature inside $\Omega(t)$ is determined.

REMARK 3.13. While the results are still rather incomplete, an interesting study of numerical approximations of solutions of (V) has been initiated by MacCamy and Weiss [61] where other references to numerical literature may be found.

3. BOUNDEDNESS AND ASYMPTOTIC BEHAVIOR BY ENERGY METHODS

3.1. Introduction

The purpose of Section 3 is to discuss the boundedness and asymptotic behavior as $t \to \infty$ of solutions of the nonlinear Volterra equation

$$u(t) + (b*Au)(t) \ni f(t) \qquad (0 \leq t < \infty) \tag{V}$$

The setting for (V) is $b : [0,\infty) \to R$ is a given kernel, A is a (possibly multivalued) maximal monotone operator on a real Hilbert space H, and $f : [0,\infty) \to H$ is a given function. The exposition is largely based on a forthcoming paper by Clément, MacCamy, and Nohel [20].

The following *general assumptions* will be assumed throughout:

$$b(t) = b_\infty + B(t), \qquad b(0) > 0, \qquad b_\infty \geq 0, \qquad B, B' \in L^1(0,\infty); \tag{H_b}$$

$$\text{A maximal monotone on H}; \tag{H_m}$$

$$f(t) = f_\infty + F(t), \qquad F \in W^{1,2}_{loc}([0,\infty;H)), \qquad F' \in L^2(0,\infty;H),$$

$$f_\infty \in H \tag{H_f}$$

here $' = d/dt$, H is a real Hilbert space with scalar product $(\cdot,\cdot)$ and norm $|\cdot|$, and W denotes the usual Sobolev space. The special case of (H_m):

$$\begin{cases} A = \partial\phi, \quad \text{where the function} \quad \phi : H \to (-\infty,\infty] \quad \text{is convex,} \\ \text{lower semicontinuous, and proper} \end{cases} \tag{H_ϕ}$$

will also play an important role in the theory. For definitions and standard results concerning maximal monotone operators and the special case of a subdifferential the reader is referred to Brézis [14].

We remark that if one adds the assumption $B' \in BV_{loc}[0,\infty)$ to assumptions (H_b), (H_m), and (H_f), then by Theorem 3.8 of Section 2.3. (V) has a unique generalized solution $u \in C([0,\infty); \overline{D(A)})$, provided $f(0) \in \overline{D(A)}$; if also $F' \in BV_{loc}[0,\infty;H)$ and $f(0) \in D(A)$, then u is a strong solution of (V) on $[0,\infty)$. If in place of (H_m) assumption (H_ϕ) is satisfied, and if $f(0) \in \overline{D(\phi)}$, then by Theorem 3.10 of Section 2, (V) has a unique strong solution u on $[0,\infty)$ such that $tu' \in L^2_{loc}(0,\infty;H)$; if $f(0) \in D(\phi)$, then $u' \in L^2_{loc}(0,\infty;H)$.

The results on boundedness and asymptotic behavior of solutions of (V) will be derived from a priori estimates obtained from the equivalent differentiated form of (V):

$$\frac{du}{dt} + b(0)au + B'*Au \ni F' \qquad (0 < t < \infty), \qquad u(0) = f(0) \qquad (V')$$

We shall distinguish two cases: (i) A satisfies assumption (H_m) and (ii) $A = \partial\phi$ with ϕ satisfying (H_ϕ). Case (i) is developed in Section 3.2, while Case (ii) is treated in Section 3.3; in each case a different energy method is used to deduce suitable a priori estimates under appropriate additional assumptions on the kernel b and the forcing term f. A number of examples illustrating each situation is presented. In particular, the theory developed in Section 3.3 is used in Section 3.4 to analyse the boundedness and asymptotic behavior of solutions of the heat flow problem (2.3), Section 1, under the physically reasonable assumptions motivated in Section 1.2.

The reader should note that if Assumption (H_ϕ) is satisfied, and if u is the strong solution of (V) of Theorem 3.10, Section 2, then (see Brézis [14, Lemma 3.3]) $\phi(u(t))$ is absolutely continuous and one has the "chain rule":

$$\frac{d}{dt}\,\phi(u(t)) = (w, \frac{du}{dt}) \qquad (w \in \partial\phi(u)(t))$$

Thus a plausible energy method for the case $A = \partial\phi$ consists of taking the scalar product of (V') by $w \in \partial\phi(u(t))$ for any solution u and integrating over an arbitrary interval (0,T). Indeed, this method is used in Section 3.3. Unfortunately, there is no analogue for the chain rule when A is a maximal monotone operator, but $A \neq \partial\phi$ for some proper, convex, l.s.c. function ϕ. For this reason the development of the theory in Section 3.2 is less direct in that the a priori estimates are derived from an equivalent equation to (V') resulting essentially from applying Proposition 3.1 and Remark 3.6 (iii), Section 2, to (V'), and then using a different energy method to obtain the a priori estimates.

It should be noted that Corollaries 2.4, 2.5, and Theorem 2.6 of Section 3.2 may be viewed as natural generalizations to Hilbert space of earlier results of Levin [51] and Londen [53] which describe the limiting behavior as $t \to \infty$ of solutions of (V) in the scalar case in which the operator A is a real function.

3.2. Boundedness and Asymptotic Properties when A is Maximal Monotone

Throughout this section we assume that assumptions (H_b), (H_m), and (H_f) are satisfied and that u is a strong or generalized solution of the Volterra equation (V) on $[0,\infty)$. As explained in Section 2 (V) is equivalent to the Cauchy problem

$$\frac{du}{dt} + b(0)Au + B'*Au \ni F' \qquad (0 < t < \infty), \qquad u(0) = f(0) \tag{V'}$$

Let k be the resolvent kernel associated with B', defined to be the unique solution of the linear Volterra equation

$$b(0)k(t) + (B'*k)(t) = \frac{-B'(t)}{b(0)} \quad \text{a.e. for} \quad 0 \leq t < \infty \tag{k}$$

by standard results, Miller [63], assumption (H_b) implies that $k \in L^1_{loc}(0,\infty)$.

We now use the method of Proposition 3.1 and Remark 3.6 (iii) of Section 2 to transform (V'). Regarding (V') as a "linear" equation for Au, the variation of constants formula for Volterra equations [63] and an integration by parts show that (V') (and hence also (V)) is equivalent to the Cauchy problem

$$\frac{1}{b(0)}\frac{du}{dt} + \frac{d}{dt}(k*u) + Au \ni f_1 \qquad (0 < t < \infty), \qquad u(0) = f(0) \tag{2.1}$$

where $f_1 : [0,\infty) \to H$ is the function given by either

$$f_1(t) = \frac{1}{b(0)} F'(t) + f(0)k(t) + (k*F')(t) \qquad (0 \leq t < \infty) \tag{2.2}$$

or

$$f_1(t) = \frac{1}{b(0)} F'(t) + k(0)f(t) + (k'*f)(t) \qquad (0 \leq t < \infty) \tag{2.3}$$

We shall use an energy method based on taking the scalar product of (2.1) by u, and also by $\sqrt{t}\,u$, and we obtain a priori estimates by integrating over an arbitrary interval $[0,T]$. We will first state the general result for (2.1) and then interpret it for (V).

THEOREM 2.1. Let u be a strong or generalized solution of the Cauchy problem (2.1) on $(0,\infty)$. Let $T > 0$ be given and let there exist constants $\varepsilon, \eta \in R$ such that

$$\text{if} \quad v \in Au, \quad \text{then} \quad \int_0^T (v,u)dt \geq \varepsilon \int_0^T |u|^2 dt \,(u \in D(A)) \tag{2.4}$$

for every $w \in L^2(0,T;H)$

$$\int_0^T \left(w(t), \frac{d}{dt}(k*w)(t)\right)dt \geq \eta \int_0^T |w|^2 dt \tag{2.5}$$

$$\varepsilon + \eta > 0 \tag{2.6}$$

(a) If $f_1 \in L^2(0,\infty;H)$, then $u \in L^\infty(0,\infty;H) \cap L^2(0,\infty;H)$;

(b) If also $\sqrt{t}\, k' \in L^1(0,\infty)$ and $\sqrt{t}\, f_1 \in L^2(0,\infty;H)$, then $\sqrt{t}\, u \in L^\infty(0,\infty;H) \cap L^2(0,\infty;H)$.

Consequently, $|u(t)| = O(\frac{1}{\sqrt{t}})$ as $t \to \infty$ and $u(t) \to 0$ strongly as $t \to \infty$.

We remark that no claim is made that the rate $|u(t)| = O(\frac{1}{\sqrt{t}})$ as $t \to \infty$ is optimal.

The coercivity assumption (2.4) concerning the maximal monotone operator A is natural for the problem in light of comparable assumptions in evolution equations. Assumption (2.5) and the hypotheses concerning k, k' will be justified in Lemmas 2.2, 2.3, below. Two different classes of kernels b in (V) are considered, each of which lead to the energy inequality (2.5), the first with $\eta = 0$, the second with $\eta > 0$, and for each of which $\sqrt{t}\, k' \in L^1(0,\infty)$. These technical lemmas, together with appropriate assumptions on the forcing function f in (V), permit an easy interpretation of Theorem 2.1 for solutions of (V). This will be done in Corollaries 2.4 and 2.5 below. The proof of Lemma 2.2 appears in the Appendix. Lemma 2.3 is an extension of a result of MacCamy [59] which in its present form was recently established by M. Tangredi [82].

LEMMA 2.2. (a) Let b satisfy assumption (H_b) with $b_\infty > 0$, and let b satisfy the frequency domain condition

(F) there exists $\delta > 0$ such that $b_\infty + \inf_{\eta \in R} [-\eta \operatorname{Im} \hat{B}(i\eta)] \geq \delta$, where $\hat{B}(i\eta) = \int_0^\infty \exp(-i\eta t)B(t)dt$. Then the resolvent kernel k of B' satisfies $k \in L^1(0,\infty)$.

(b) If also $B' \in L^2(0,\infty)$, then $k \in L^2(0,\infty)$; if also $B'' \in L^1(0,\infty)$, then $k' \in L^1(0,\infty)$.

(c) If the assumptions of (a) are satisfied, $B'' \in L^1(0,\infty)$, and B is a kernel of positive type on $[0,\infty)$, then for every $T > 0$ and for every $w \in L^2(0,T)$

$$\int_0^T w(t) \frac{d}{dt}(k*w)(t)dt \geq 0$$

(d) If the assumptions of (a) and (b) are satisfied, and $\sqrt{t}\,B' \in L^1(0,\infty) \cap L^2(0,\infty)$, $\sqrt{t}\,B'' \in L^1(0,\infty)$, then $\sqrt{t}\,k \in L^1(0,\infty) \cap L^2(0,\infty)$, and $\sqrt{t}\,k' \in L^1(0,\infty)$.

LEMMA 2.3. Let b satisfy assumption (H_b) with $b_\infty = 0$, and let

(i) $t^j B^{(m)} \in L^1(0,\infty)$ $(j = 0,1,2,3;\ m = 0,1,2,3)$,

(ii) B be strongly positive on $[0,\infty)$.

Let k be the resolvent kernel of B'. Then:

(a) $k \in C^1[0,\infty)$;

(b) $k(t) = k_\infty + K(t)$, $k_\infty = (\int_0^\infty B(t)dt)^{-1} > 0$, $K^{(m)} \in L^1(0,\infty)$ $(m = 0,1,2)$;

(c) if also $B, B', \sqrt{t}\,B, \sqrt{t}\,B' \in L^2(0,\infty)$ one has $K, \sqrt{t}\,K \in L^2(0,\infty)$;

(d) for every $T > 0$ and for every $w \in L^2(0,T)$ there exists $\eta > 0$ such that

$$\int_0^T w(t) \frac{d}{dt} (k*w)(t)dt \geq \eta \int_0^T |w(t)|^2 dt$$

(e) if assumptions (i) hold for $j,m = 0,1,2,3$, $t^4 B \in L^1(0,\infty)$, and assumptions (ii) hold, one has $\sqrt{t}\,k' \in L^1(0,\infty)$.

We shall mention some examples of kernels b which satisfy the assumptions of Lemmas 2.2 and 2.3.

Let

$$B : [0,\infty) \to R^+ \text{ be positive, nonincreasing, and convex} \tag{2.7}$$

and satisfy the smoothness and integrability assumptions in (H_b). Then B is a kernel of positive type on $[0,\infty)$ (see [67]), and

$$-\eta \operatorname{Im}\hat{B}(i\eta) = \eta \int_0^\infty \sin\eta t\; B(t)dt \geq 0 \qquad (\eta \in R)$$

Thus if $b_\infty > 0$ is any constant, $b(t) = b_\infty + B(t)$ satisfies the frequency domain condition (F) with $\delta = b_\infty$, and (see Lemma 2.2(a)) $k \in L^1(0,\infty)$. If, in addition, B satisfies the remaining smoothness and integrability assumptions of Lemma 2.2, all conclusions of Lemma 2.2, and assumptions (2.5) with $\eta = 0$, and $\sqrt{t}\,k' \in L^1(0,\infty)$ of Theorem 2.1 are satisfied.

Consider again B in (2.7). In addition, assume that

$$\text{the measure } dB' \text{ has a nonzero absolutely continuous part;} \tag{2.8}$$

then (see [67, Corollary 2.2]), B is strongly positive on $[0,\infty)$ (for example, $B \in C[0,\infty)$, $(-1)^k B^{(k)}(t) \geq 0$, $0 < t < \infty$, $k = 0,1,2$, $B'(t) \not\equiv 0$).

Thus if B satisfies (2.7), (2.8), and the integrability and smoothness assumptions of Lemma 2.3, and if $b(t) \equiv B(t)$ ($b_\infty = 0$), then all conclusions of Lemma 2.3, assumptions (2.5) with $\eta > 0$, and $\sqrt{t}\, k' \in L^1(0,\infty)$ of Theorem 2.1 are satisfied.

Next, consider

$$B(t) = \sum_{j=1}^{m} B_j e^{-\lambda_j t} \cos w_j t \qquad (B_j \geq 0,\ \lambda_j \geq 0,\ w_j \in R) \tag{2.9}$$

with strict inequalities holding for at least one j (if $w_j = 0$, $j = 1,\ldots, m$, B satisfies both (2.7) and (2.8)). This function B is strongly positive on $[0,\infty)$ (see [67]), since by direct calculation

$$\operatorname{Re} \hat{B}(i\eta) = \frac{1}{2} \sum_{i=1}^{m} B_j \lambda_j \left(\frac{1}{\lambda_j^2 + (\eta - w_j)^2} + \frac{1}{\lambda_j^2 + (\eta + w_j)^2}\right) \qquad (\eta \in R)$$

Moreover, B satisfies all other assumptions of Lemma 2.3. Thus if $b(t) = B(t)$ ($b_\infty = 0$), all conclusions of Lemma 2.3, assumptions (2.5) with $\eta > 0$ and $\sqrt{t}\, k' \in L^1(0,\infty)$ of Theorem 2.1 are satisfied.

For the kernel B in (2.9) one has

$$-\eta \operatorname{Im} \hat{B}(i\eta) = \sum_{i=1}^{m} B_j \frac{\eta^2(\eta^2 + \lambda_j^2 - w_j^2)}{(\lambda_j^2 + w_j^2 - \eta^2)^2 + 4\lambda_j^2\eta^2} \qquad (\eta \in R)$$

Thus $b(t) = b_\infty + B(t)$, where $b_\infty > 0$ is any constant, satisfies the frequency domain condition (F) of Lemma 2.2 if $\lambda_j \geq w_j$ ($j = 1,\ldots,m$). Evidently, b is a kernel of positive type on $[0,\infty)$. Therefore, if $b(t) = b_\infty + B(t)$, $b_\infty > 0$, B defined by (2.9) with $\lambda_j \geq w_j$ ($j = 1,\ldots,m$), all conclusions of Lemma 2.2 (but not of Lemma 2.3), assumptions (2.5) with $\eta = 0$ and $\sqrt{t}\, k' \in L^1(0,\infty)$ of Theorem 2.1 are satisfied.

Incidentally, if $b_\infty > 0$ is any constant, and if

$$B(t) = \sum_{i=1}^{m} B_j e^{-\lambda_j t} \sin w_j t \qquad (\lambda_j \geq 0,\ w_j \geq 0) \tag{2.10}$$

with strict inequalities holding for at least one j, then the frequency domain condition (F) of Lemma 2.2 is satisfied with $\delta = b_\infty$. However, such a kernel b is not of positive type.

Lemma 2.2 combined with appropriate assumptions on A and f yield the following easy interpretation of Theorem 2.1 for solutions of (V).

COROLLARY 2.4. Let assumptions (H_b) with $b_\infty > 0$, (H_m) and (H_f) with f_∞ arbitrary be satisfied. In addition, assume that b satisfies the hypotheses of Lemma 2.2, and that $\sqrt{t}\, F' \in L^2(0,\infty;H)$. Let u be a strong or generalized solution of (V) on $[0,\infty)$. If the coercivity assumption (2.4) holds with $\varepsilon > 0$, then u and $\sqrt{t}\, u \in L^\infty(0,\infty;H) \cap L^2(0,\infty;H)$ and $u(t) \to 0$ strongly as $t \to \infty$.

Indeed, define f_1 by (2.2). By (H_f) and Lemma 2.2 ($k \in L^1(0,\infty) \cap L^2(0,\infty)$, $k' \in L^1(0,\infty)$) one trivially has $f_1 \in L^2(0,\infty;H)$. By Lemma 2.2 one also has $\sqrt{t}\, k \in L^1(0,\infty) \cap L^2(0,\infty)$ and $\sqrt{t}\, k' \in L^1(0,\infty)$. These together with the assumption $\sqrt{t}\, F' \in L^2(0,\infty;H)$ used in (2.2) show that $\sqrt{t}\, f_1 \in L^2(0,\infty;H)$; the fact that $\sqrt{t}\, (k*F') \in L^2(0,\infty;H)$ in (2.2) follows from the straightforward estimate

$$\int_0^T t\left|\int_0^t k(t-s)F'(s)ds\right|^2 dt \le 2\|k\|^2_{L^1(0,\infty)} \|\sqrt{t}\, F'\|^2_{L^2(0,\infty;H)} + 2\|\sqrt{t}\, k\|^2_{L^1(0,\infty)} \|F'\|^2_{L^2(0,\infty;H)} \quad \text{(for all } T > 0\text{)}$$

By Lemma 2.2 again, (2.5) holds with $\eta = 0$. Thus if $\varepsilon > 0$ in (2.4), the result of Corollary 2.4 follows by applying Theorem 2.1.

Lemma 2.3 combined with appropriate assumptions on A and f yield a different interpretation of Theorem 2.1 for solutions of (V).

COROLLARY 2.5. Let assumptions (H_b) with $b_\infty = 0$, (H_m) and (H_f) with $f_\infty = 0$ be satisfied. In addition, assume that $b(t) = B(t)$ satisfies the assumptions of Lemma 2.3, and that $f \equiv F$ also satisfies $f, \sqrt{t}\, f, \sqrt{t}\, f' \in L^2(0,\infty;H)$. Let u be a strong or generalized solution of (V) on $[0,\infty)$. If the operator A satisfies the coercivity assumption (2.4) with $\varepsilon \geq 0$ (or even $\varepsilon > -\eta$, where $\eta > 0$ is the constant in Lemma (2.3d), then u and $\sqrt{t}\, u \in L^\infty(0,\infty;H) \cap L^2(0,\infty;H)$, and $u(t) \to 0$ strongly as $t \to \infty$.

The proof of Corollary 2.5 is similar to that of Corollary 2.4, except that f_1 must now be defined by (2.3), and Lemma 2.3 is used in place of Lemma 2.2. Note also that the additional assumptions concerning f, $\sqrt{t}\, f$ are essential.

The important case $b_\infty = 0$ in (H_b), $b \equiv B$ satisfying the assumptions of Lemma 2.3, and $f_\infty \neq 0$ in (H_f) is not covered by Corollary 2.5. In this situation Theorem 2.1 must be modified in the following manner.

THEOREM 2.6. Let the assumptions (H_b) $(b_\infty = 0)$, (H_m), (H_f) with f_∞ arbitrary, and the assumptions of Lemma 2.3 be satisfied. In addition, assume that F, $\sqrt{t}$ F, $\sqrt{t}$ f' $\in L^2(0,\infty;H)$. Let u be a strong of generalized solution of (V) on $[0,\infty)$, let u_∞ be the unique solution of the limit equation corresponding to (V):

$$u_\infty + (\int_0^\infty B(t)dt)Au_\infty \ni f_\infty \tag{V_L}$$

Let the operator A satisfy the coercivity condition:

$$\begin{cases} \text{if} \quad v \in Au \quad \text{and} \quad v_\infty \in Au_\infty \quad \text{and} \quad T > 0, \quad \text{then} \\ \int_0^T (v(t) - v_\infty, u(t) - u_\infty)dt \geq \varepsilon \int_0^T |u(t) - u_\infty|^2 dt \\ \text{for some } \varepsilon \geq 0 \quad (\varepsilon > -\eta \text{ is sufficient; see Lemma 2.3d}) \end{cases} \tag{2.11}$$

Then $u - u_\infty$ and $\sqrt{t}\,(u - u_\infty) \in L^\infty(0,\infty;H) \cap L^2(0,\infty;H)$; consequently $u(t) \to u_\infty$ strongly as $t \to \infty$ and $|u(t) - u_\infty| = O(\frac{1}{\sqrt{t}})$ as $t \to \infty$.

REMARK 2.7. Since $b \equiv B$ satisfies the hypothesis of Lemma 2.3, B is strongly positive on $[0,\infty)$, and therefore $\int_0^\infty B(t)dt > 0$. Since the operator A is maximum monotone on H the limit equation (V_L) has a unique solution for any $f_\infty \in H$; in particular, if $f_\infty = 0$, $u_\infty = 0$ and in this case Theorem 2.6 reduces to Corollary 2.5.

Corollaries 2.4, 2.5 and Theorem 2.6 together form the natural generalization to Hilbert space of corresponding scalar results for (V) due to Levin [51] and Londen [53].

Sketch of Proof of Theorem 2.1. (a) Take the scalar product of (2.1) with u and integrate from 0 to T. Using (2.4) and (2.5) we obtain

$$\frac{1}{2b(0)} |u(T)|^2 + (\eta + \varepsilon) \int_0^T |u(t)|^2 dt \leq \frac{1}{2b(0)} |f(0)|^2 + \int_0^T (f_1(t),u(t))dt$$

Since $\eta + \varepsilon > 0$, and $f_1 \in L^2(0,\infty;H)$, the assertion (a) follows by standard estimates.

(b) Next take the scalar product of (2.1) with tu and integrate from 0 to T. An integration by parts yields

$$\begin{aligned} &\frac{T}{2b(0)} |u(t)|^2 + \int_0^T t(u(t), \quad \frac{d}{dt} (k*u)(t))dt + \varepsilon \int_0^T t|u(t)|^2 dt \\ &\leq \frac{1}{2b(0)} \int_0^T |u(t)|^2 dt + \int_0^T (tf_1(t), \quad u(t))dt \end{aligned} \tag{2.12}$$

A straightforward calculation shows that (see [20])

$$\int_0^T t(u(t), \quad \frac{d}{dt} (k*u)(t))dt = I + J$$

where

$$I = \int_0^T \left(\sqrt{t}\, u(t),\quad \frac{d}{dt}(k*\sqrt{t}\, u)(t)\right)dt \geq \eta \int_0^T t|u(t)|^2 dt$$

and where

$$J = \int_0^T \left(\sqrt{t}\, u(t),\quad \int_0^t k'(t-\tau)(\sqrt{t}-\sqrt{\tau})u(\tau)d\tau\right)dt$$

Using $\sqrt{t} - \sqrt{\tau} < \sqrt{t-\tau}$, $0 \leq \tau \leq t$, $u \in L^2(0,\infty;H)$, and $\sqrt{t}\, k' \in L^1(0,\infty)$ yields

$$|J| \leq \left(\int_0^T t|u(t)|^2 dt\right)^{1/2} \| u \|_{L^2(0,\infty;H)} \int_0^\infty \sqrt{t}|k'(t)|dt$$

Substitution of the estimates for I, J in (2.12) gives the final inequality

$$\frac{T}{2b(0)} |u(t)|^2 + (\eta + \varepsilon) \int_0^T t|u(t)|^2 dt \leq \frac{1}{2b(0)} \| u \|^2_{L^2(0,\infty;H)}$$

$$+ \| u \|_{L^2(0,\infty;H)} \| \sqrt{t}\, k' \|_{L^1(0,\infty)} \left(\int_0^T t|u(t)|^2 dt\right)^{1/2}$$

$$+ \| \sqrt{t}\, f_1 \|_{L^2(0,\infty;H)} \| u \|_{L^2(0,\infty;H)}$$

The conclusion $\sqrt{t}\, u \in L^\infty(0,\infty;H) \cap L^2(0,\infty;H)$ follows by standard estimates using $\varepsilon + \eta > 0$, $u \in L^2(0,\infty;H)$ by (a), and the assumptions $\sqrt{t}\, k' \in L^1(0,\infty)$ and $\sqrt{t}\, f_1 \in L^2(0,\infty;H)$. This completes the sketch of the proof.

Proof of Theorem 2.6. The proof will be reduced to that of Theorem 2.1 by the following steps. First by Lemma 2.3 $\int_0^\infty B(t)dt = k_\infty^{-1} > 0$. Therefore, the limit equation (V_L) can be written in the form

$$k_\infty u_\infty + Au_\infty \ni k_\infty f_\infty$$

which is the same as

$$\frac{1}{b(0)} \frac{d}{dt} u_\infty + \frac{d}{dt}(k*u_\infty) + Au_\infty \ni k_\infty f_\infty + (k(t) - k_\infty)u_\infty \tag{2.13}$$

Next, subtracting (2.13) from (2.1) gives

$$\frac{1}{b(0)} \frac{d}{dt}(u - u_\infty) + \frac{d}{dt} k*(u - u_\infty) + Au - Au_\infty \ni F_1(t) \tag{2.14}$$

$$(0 < t < \infty)$$

where by an elementary calculation

$$F_1(t) = \frac{1}{b(0)} F'(t) + k(0)F(t) + K(t)f_\infty + (k'*F)(t) - u_\infty K(t) \tag{2.15}$$

Lemma 2.3 and the assumptions concerning F clearly imply that F_1 satisfies the same assumptions as f_1 in Theorem 2.1. The method of proof of Theorem 2.1 applied to (2.14), (2.15), where the coercivity assumption (2.11) is used in place of (2.4), now yields the needed a priori estimates for $u - u_\infty$ and $\sqrt{t}\,(u - u_\infty)$, and completes the proof.

EXAMPLE 2.8. We give an example of a maximum monotone operator A in (V) which is not a subdifferential, and for which the theory developed in this section is applicable. Let $\Omega \subset R^n$ be a bounded open set with smooth boundary $\partial\Omega$. Let H be the Hilbert space $L^2(\Omega)$. Let β be a maximum monotone graph with $0 \in \beta(0)$ and with primitive j (i.e. $\beta = \partial j$). Let A_1 be the operator defined by

$$D(A_1) = \{u : u \in H_1^0(\Omega) \cap H^2(\Omega), \qquad \beta(u) \in L^2(\Omega)\}$$

$$A_1 u = -\Delta u + \beta(u) \qquad (u \in D(A_1))$$

It is clear [15] that A_1 is maximum monotone on H since $A_1 = \partial\phi_1$, where $\phi_1 : H \to (-\infty,\infty]$ is the proper, convex, l.s.c. function given by

$$\phi_1(u) = \begin{cases} \frac{1}{2}\int_\Omega |\nabla u|^2 dx + \int_\Omega j(u)dx & \text{if} \quad u \in H_0^1(\Omega) \\ \quad \text{and} \quad j(u) \in L^1(\Omega) & \\ +\infty \quad \text{otherwise} & \end{cases}$$

Define $L(u) = \sum_{i=1}^{n} b_i\, \partial u/\partial x_i$ $(b_i \in R,\ u \in H_0^1(\Omega))$. By the divergence theorem L(u) is monotone and $(u,L(u)) = \int_\Omega uL(u)dx = 0$. Finally, following Pazy [70, Ex. 3.5] define

$$A = A_1 + L$$

By a perturbation theorem of Crandall and Pazy [27], A is maximum monotone on $H = L^2(\Omega)$, and by an easy calculation using Green's theorem and the Poincaré inequality there exists a constant $\varepsilon > 0$ such that

$$(Au,u) = (A_1u,u) = -\int_\Omega u\Delta u dx + \int_\Omega u\beta(u)dx \geq \int_\Omega |\nabla u|^2 dx$$

$$\geq \varepsilon \| u \|^2_{L^2(\Omega)}$$

Thus A satisfies the coercivity assumption (2.4) for every $T > 0$.

REMARK 2.9. The concept of strong positivity of a kernel plays an important role in stability theory for Volterra equations (see Halanay [40], Nohel and Shea [67], Staffans [74-78], Crandall, Londen and Nohel [28]). As we have seen kernels B satisfying (2.7) and (2.8), as well as oscillatory kernels B of the form (2.9) are strongly positive on $[0,\infty)$. If $b(t) = b_\infty + B(t)$ where $b_\infty \geq 0$ and B is strongly positive on $[0,\infty)$ one can extract some information about the behavior of solutions of (V) as $t \to \infty$.

In particular, one can establish $u \in L^2(0,\infty;H)$ (Theorem 2.1 part (a)) by another energy method directly from (V):

PROPOSITION 2.10. Let the general assumptions (H_b), (H_m), (H_f) be satisfied. Let u be a strong or generalized solution of (V) on $[0,\infty)$. If $F \in L^2(0,\infty;H)$, if B is strongly positive on $[0,\infty)$, if the coercivity assumption (2.4) is satisfied with $\varepsilon > 0$, and if $f_\infty = 0$ whenever $b_\infty = 0$, then $u \in L^2(0,\infty;H)$.

Proof. Let $v \in Au$ and let $0 < T < \infty$ be given. Take the scalar product of (V) by v and integrate from 0 to T obtaining

$$\begin{aligned} &\int_0^T (v(t), u(t))dt + \frac{b_\infty}{2} \left|\int_0^T v(t)dt\right|^2 + Q_B[v;T] \\ &\quad \leq \int_0^T (v(t), f(t))dt \end{aligned} \tag{2.16}$$

where

$$Q_B[v;T] = \int_0^T (v(t), (B*v)(t))dt$$

Since $B \in L^1(0,\infty)$ is strongly positive and $F, F' \in L^1(0,\infty;H)$, a result of Staffans [80, Proposition 4.1] shows that there exists a constant $\gamma > 0$ such that

$$\left|\int_0^T (v(t), F(t))dt\right| \leq \gamma\{Q_B[v;T]\}^{1/2}$$

Using this, the coercivity assumption (2.4) $(\varepsilon > 0)$, and the obvious estimate

$$\left|\int_0^T (v(t), f_\infty)dt\right| \leq \frac{b_\infty}{4} \left|\int_0^T v(t)dt\right|^2 + \frac{1}{b_\infty} |f_\infty|^2 \qquad (b_\infty > 0)$$

in (2.16) yields the inequalities

$$\begin{aligned} &\varepsilon \int_0^T |u(t)|^2 dt + \frac{b_\infty}{4} \left|\int_0^T v(t)dt\right|^2 \\ &\quad + Q_B[v;t] \leq \gamma\{Q_B[v;T]\}^{1/2} + \frac{|f_\infty|^2}{b_\infty} \qquad (b_\infty > 0) \end{aligned} \tag{2.17}$$

and

$$\begin{aligned} &\varepsilon \int_0^T |u(t)|^2 dt + Q_B[v;T] \leq \gamma\{Q_b[v;T]\}^{1/2} \\ &(b_\infty = 0,\ f_\infty = 0) \end{aligned} \tag{2.18}$$

The result of Proposition 2.10 now follows from (2.17) and (2.18) by completing the square on the Q_B terms.

Incidentally, we have also shown that

$$\sup_{T>0} Q_B[v;T] < \infty \qquad (v \in Au) \tag{2.19}$$

and if $b_\infty > 0$, also that

$$\sup_{T>0} \left|\int_0^T v(t)dt\right| < \infty$$

It follows from (2.19) using another result of Staffans (see Crandall, Nohel, Londen [28, Lemma 3.1] in a Hilbert space setting) that

$$\sup_{T>0} |(B*v)(T)| < \infty \qquad (v \in Au)$$

Unfortunately, there appears to be no direct way to establish also that u(t) is uniformly continuous on $[0,\infty)$ if assumption (H_m) is satisfied and $A \neq \partial\phi$ (if $A = \partial\phi$ this can be done as in Theorem 3.1 below). The uniform continuity together with $u \in L^2(0,\infty;H)$ would imply that $u \to 0$ as $t \to \infty$ strongly in H. This provides at least one motivation for the indirect method of Theorem 2.1.

3.3. Boundedness and Asymptotic Properties When $A = \partial\phi$

Let the general assumptions (H_b), (H_ϕ), (H_f) be satisfied and let u be a strong or generalized solution of (V) or $[0,\infty)$. In this section we shall obtain different boundedness and asymptotic results for the case $A = \partial\phi$, and when $b_\infty > 0$ in (H_b). These results, which are motivated by the physical problem discussed in Section 1.2 are deduced from a priori estimates which are obtained directly from the equivalent Cauchy problem;

$$\frac{du}{dt} + b(0)Au + B' * Au \ni F' \qquad (0 < t < \infty), \qquad u(0) = f(0) \tag{V'}$$

THEOREM 3.1. Let the general assumptions (H_b) with $b_\infty > 0$, (H_ϕ), (H_f) be satisfied and let u be a strong or generalized solution of (V) on $[0,\infty)$. If the kernel b satisfies the frequency domain condition $(\mathcal{F})$ of Lemma 2.2 and if

$$\inf_{z\in H} \phi(z) > -\infty \tag{3.1}$$

then

$$\sup_{0<t<\infty} \phi(u(t)) < \infty \tag{3.2}$$

if $V \in \partial\phi(u)$, then

$$v \in L^2(0,\infty;H) \tag{3.3}$$

$$\frac{du}{dt} \in L^2(0,\infty;H) \tag{3.4}$$

and

$$u \text{ is strongly uniformly continuous on } [0,\infty) \tag{3.5}$$

If also $\lim_{|u|\to\infty} \phi(u) = +\infty$, then

$$\sup_{0\leq t<\infty} |u(t)| < \infty \tag{3.6}$$

and

$$\lim_{t\to\infty} \phi(u(t)) = \phi_\infty = \inf_{z\in H} \phi(z) \quad \text{exists} \tag{3.7}$$

Moreover, if the inclusion $\partial\phi(w) \ni 0$ implies $w = 0$, then

$$u(t) \to 0 \quad \text{(weakly) as} \quad t \to \infty \tag{3.8}$$

The frequency domain condition (F) is satisfied by several classes of kernels b with $b_\infty > 0$ as was seen in Section 3.2 (see examples of $b = b_\infty + B$ with B given by (2.7), (2.9), and (2.10)). Thus Theorem 3.1 generalizes a recent result of S. -O. Londen [54, Corollary 2] and a result of V. Barbu [6, Theorem 2].

The assumptions concerning ϕ in Theorem 3.1 are not sufficient to obtain strong convergence of u(t) to zero as $t \to \infty$. For this result one needs the coercivity condition (2.4) with $\varepsilon > 0$. If (2.4) is satisfied with $v \in \partial\phi(u)$ it is a standard result (see Brézis [14]) that the inclusion $\partial\phi(w) \ni 0$ has $w = 0$ as the only solution, and that $0 \in D(\partial\phi)$. Then the definition of the subdifferential [14] implies that

$$\phi(u) \geq \phi(0) \qquad (u \in H)$$

and therefore assumption (3.1) of Theorem 3.1 holds. This motivates the following results which complement Corollary 2.4 for the case $A = \partial\phi$. Note that in Theorem 3.2 below only the frequency domain condition (F), but not the assumption that B is a kernel of positive type (see Lemma 2.2), is needed. Also note that here the assumption on F is less restrictive.

THEOREM 3.2. Let the general assumptions (H_b) with $b_\infty > 0$, (H_ϕ), (H_f) be satisfied, and let u be a strong or generalized solution of (V) on $[0,\infty)$. Let b satisfy the frequency domain condition (F), and for $v \in \partial\phi(u)$ let the coercivity condition (2.4) with $\varepsilon > 0$ be satisfied. Then conclusions

(3.2)-(3.5) of Theorem 3.1 hold, and $u \in L^2(0,\infty;H)$, which implies that $u(t) \to 0$ strongly as $t \to \infty$.

REMARK 3.3. If $b(t) \equiv b_\infty > 0$ in (V), a case not excluded in Theorems 3.1 and 3.2, the above theorem and its proof yield a simple boundedness and asymptotic behavior result for the evolution equation

$$\frac{du}{dt} + b_\infty \partial\phi(u) \ni g, \qquad u(0) = u_0$$

where $g = F'$; compare Brézis [14, Theorem 3.11] where $g \in L^1(0,\infty;H)$.

REMARK 3.4. If the coercivity condition (2.4) with $\varepsilon > 0$ and $A = \partial\phi$ is replaced by the more general condition: for every $T > 0$ there exists $\varepsilon > 0$ such that

$$\begin{aligned} \text{if} \quad v \in \phi(u), \quad \text{then} \quad & \int_0^T (v(t), u(t) - z)dt \\ & \geq \int_0^T |u(t) - z|^2 dt \end{aligned} \tag{2.4'}$$

for some $z \in H$, then it is easy to show that the inclusion $\partial\phi(w) \ni 0$ has $w = z$ as the only solution, and that

$$\phi(u) \geq \phi(z) \qquad (u \in H)$$

Then the method of proof of Theorem 3.2 easily yields that $u(t) \to z$ strongly as $t \to \infty$.

REMARK 3.5. In Theorems 3.1 and 3.2 the assumption $b_\infty > 0$ in (H_b) is crucial; if $b_\infty = 0$ the frequency domain condition (F) cannot be satisfied (see examples (2.7), (2.9), and (2.10)) for any $\delta > 0$. On the other hand, in these theorems f_∞ in (H_f) is arbitrary and the case $f_\infty = 0$ is not ruled out, provided $b_\infty > 0$. If $b_\infty = 0$ in (H_b), one can, of course, still apply Corollary 2.5 if $f_\infty = 0$, and Theorem 2.6 if $f_\infty \neq 0$, with $A = \partial\phi$.

Proof of Theorem 3.1. (a) Let $0 < T < \infty$ be arbitrary; take the scalar product of (V') with $v \in \partial\phi(u)$ and integrate over $[0,T]$. Using $d/dt\ \phi(u(t)) = (v(t), du/dt\ (t))$ a.e. (see Brézis [14]) one obtains

$$\begin{aligned} & \phi(u(T)) + b(0) \int_0^T |v(t)|^2 dt + Q_{B'}[v;T] \\ & = \int_0^T (F'(t), v(t))dt + \phi(f(0)) \end{aligned} \tag{3.9}$$

where

$$Q_{B'}[v;T] = \int_0^T (v(t), B' * v(t))dt$$

We next apply a frequency domain method (see Nohel and Shea [67]) to $Q_{B'}$. Define $v_T(t) = v(t)\chi[0,T]$ and its Fourier transform

$$\tilde{v}_T(\eta) = \int_{-\infty}^{\infty} e^{-i\eta t} v_T(t)dt$$

Extend B' evenly to $(-\infty,0)$ by $B'(-t) = B'(t)$ $(0 < t < \infty)$. In the following calculation use is made of the hypothesis $B' \in L^1(0,\infty)$ and the Parseval and convolution theorems:

$$Q_{B'}[v;T] = \int_0^T (v(t), B'*v(t))dt = \frac{1}{2}\int_0^T (v(t), \int_0^T B'(t-\tau)v(\tau)d\tau)dt$$
$$= \frac{1}{2}\int_{-\infty}^{\infty} (v_T(t), \int_{-\infty}^{\infty} B'(t-\tau)v_T(\tau)d\tau)dt$$
$$= \frac{1}{4\pi}\int_{-\infty}^{\infty} |\tilde{v}_T(\eta)|^2 \tilde{B}'(\eta)d\eta$$

Since B' is even, $\tilde{B}'(\eta) = 2\mathrm{Re}\,\hat{B}'(i\eta)$ $(\eta \in R)$, where $\hat{\cdot}$ denotes the Laplace transform. The assumptions $B, B' \in L^1(0,\infty)$ and the familiar formula $\hat{B}'(i\eta) = i\eta\,\hat{B}(i\eta) - B(0)$ yield $\mathrm{Re}\,\hat{B}'(i\eta) = -\eta\mathrm{Im}\,\hat{B}(i\eta) - B(0)$. Therefore,

$$Q_{B'}[v;T] = \frac{1}{2\pi}\int_{-\infty}^{\infty} |\tilde{v}_T(\eta)|^2[-\eta\mathrm{Im}\,\hat{B}(i\eta) - B(0)]d\eta$$

Substituting this result into (3.9) and using $\int_0^T |v(t)|^2dt = 1/2\pi \int_{-\infty}^{\infty} |\tilde{v}_T(\eta)|^2d\eta$, as well as $b(0) - B(0) = b_\infty$, the frequency domain condition (F), and Parseval's theorem again, yields

$$\phi(u(T)) + \delta\int_0^T |v(t)|^2dt \le |\phi(f(0))| + \int_0^T |(F'(t), v(t))|dt \qquad (0 \le T < \infty) \tag{3.10}$$

The assumption $F' \in L^2(0,\infty)$, Cauchy-Schwarz and an elementary inequality give the estimate

$$\phi(u(T)) + \frac{\delta}{2}\int_0^T |v(t)|^2dt \le |\phi(f(0))| + \frac{1}{2\delta}\int_0^\infty |F'(t)|^2dt < \infty \qquad (0 < T < \infty) \tag{3.11}$$

Assumption (3.1) used in (3.11) yields conclusions (3.2), (3.3), and (3.6).

Returning to (V') and using $B' \in L^1(0,\infty)$, $v \in L^2(0,\infty;H)$, $F' \in L^2(0,\infty;H)$, we get conclusion (3.4). Combining (3.3) and (3.4) with $d/dt\,\phi(u(t)) = (v(t), du/dt\,(t))$ yields $d/dt\,\phi(u(t)) \in L^1(0,\infty)$, and this together with assumption (3.1) implies that $\lim_{t\to\infty} \phi(u(t))$ exists. To establish all of (3.7) we use the definition of subdifferential: for every $v \in \partial\phi(u)$ and for every $w \in H$ $\phi(u(t)) \le \phi(w) + (v(t), u(t) - w)$, $0 \le t < \infty$. Since $u \in L^\infty(0,\infty;H)$ and $v \in L^2(0,\infty;H)$ there exists a sequence $\{t_n\} \to \infty$ as

$n \to \infty$ such that $(v(t_n), u(t_n) - w) \to 0$ as $n \to \infty$; this proves (3.7), and from it easily (3.8). To prove (3.5) take $\tau < t$ and use (3.4) and Cauchy-Schwarz obtaining:

$$|u(t) - u(\tau)| \leq \int_\tau^t \left|\frac{du}{dt}(s)\right| ds \leq \sqrt{t - \tau}\, \{\int_0^\infty \left|\frac{du}{dt}(s)\right|^2 ds\}^{1/2}$$

$$\leq K\sqrt{t - \tau} \qquad (0 \leq \tau < t < \infty)$$

This completes the proof of Theorem 3.1.

Proof of Theorem 3.2. As remarked in the paragraph preceding Theorem 3.2 the coercivity condition (2.4) implies that

$$\operatorname*{Inf}_{z \in H} \phi(z) \geq \phi(0) > -\infty$$

so that assumption (3.1) is satisfied. Thus conclusions (3.2)-(3.5) follow immediately from Theorem 3.1. In view of (3.5) the conclusion $u(t) \to 0$ strongly as $t \to \infty$ follows once it is shown that $u \in L^2(0,\infty;H)$. But using assumption (2.4) with $\varepsilon > 0$ and $v \in L^2(0,\infty;H)$ for $v \in \partial\phi(u)$ (proved in (3.3)) one has

$$\varepsilon \int_0^T |u(t)|^2 dt \leq \int_0^T (v(t), u(t))dt \leq \frac{\varepsilon}{2} \int_0^T |u(t)|^2 dt + \frac{1}{2\varepsilon} \int_0^T |v(t)|^2 dt$$

Thus

$$\frac{\varepsilon}{2} \int_0^T |u(t)|^2 dt \leq \frac{1}{2\varepsilon} \int_0^\infty |v(t)|^2 dt < \infty \qquad (0 < T < \infty)$$

Since $\varepsilon > 0$, this completes the proof of Theorem 3.2.

3.4. Application to a Problem of Heat Flow in a Material with Memory

In this section we study the heat flow problem (2.3) in one space dimension formulated in Section 1.2. We use the existence and uniqueness theory of Section 2 and the boundedness and asymptotic theory of Section 3.3, to deduce the principal result, Theorem 4.5 below

For clarity of exposition we restate the heat flow problem (2.3) Section 1.2:

$$\begin{cases} \frac{\partial}{\partial t}[b_0 u + \beta * u] = c_0 \sigma(u_x)_x - \gamma * \sigma(u_x)_x + h \\ (0 < t < \infty,\ 0 < x < 1) \\ u(0,x) = u_0(x) \qquad (0 < x < 1) \\ u(t,0) = u(t,1) \equiv 0 \qquad (t \geq 0) \end{cases} \tag{4.1}$$

where subscripts denote differentiation with respect to x. We assume that the conditions (which were motivated in Section 1):

$$b_0 + \operatorname{Re} \hat{\beta}(i\eta) > 0 \qquad (\eta \in R) \tag{PW}$$

$$c_0 - \int_0^\infty \gamma(\tau)d\tau > 0 \tag{γ}$$

where $\hat{\beta}(i\eta) = \int_0^\infty \beta(t)\exp(-i\eta t)dt$, are satisfied. We also assume that the function $\sigma : R \to R$ satisfies assumptions (σ) of Lemma 2.3 (Section 1), $u_0 \in H_0^1(0,1)$, and that the external heat supply $h \in L^1(0,\infty;L^2(0,1)) \cap L^2(0,\infty;L^2(0,1))$. Under these assumptions we have seen in Section 1 that the initial-boundary value problem (4.1) is equivalent to the abstract Volterra equation

$$u + b*Au \ni f \qquad (0 \le t < \infty) \tag{V}$$

in the present application

$$b(t) = \frac{C(t)}{b_0} + (\rho*C)(t) \qquad (0 \le t < \infty) \tag{b}$$

$$f(t) = \frac{G(t,\cdot)}{b_0} + (\rho*G)(t,\cdot) \qquad (0 \le t < \infty) \tag{f}$$

where ρ is the unique solution of the resolvent equation

$$b_0\rho(t) + (\beta*\rho)(t) = -\frac{\beta(t)}{b_0} \qquad (0 \le t < \infty) \tag{ρ}$$

$$C(t) = c_0 - \int_0^t \gamma(\tau)d\tau \qquad (0 \le t < \infty)$$

$$G(t,x) = b_0u_0(x) + \int_0^t h(\tau,x)d\tau \qquad (0 \le t < \infty,\ 0 < x < 1)$$

and the nonlinear operator $A = \partial\phi$ satisfies assumption (H_ϕ) with ϕ given by the proper, convex, l.s.c. function defined in and satisfying the properties of Lemma 2.3 in Section 1. We recall that in two or three space dimensions the corresponding heat flow problem in a bounded domain Ω with smooth boundary Γ also satisfies the Volterra equation (V) with b and f as above, but with $L^2(0,1)$ replaced by $L^2(\Omega)$, and with $A = \partial\phi$ where ϕ is the proper convex, l.s.c. function defined in and satisfying the properties of Remark 2.4 in Section 1. We recall also that the key properties of the kernel b and of the forcing term f for the heat flow problem are stated in Lemmas 2.1 and 2.2 of Section 1. With these properties in mind all that is needed in order to apply Theorems 3.1 and 3.2 of Section 3 to the problem under study is to show that the frequency domain assumption (F) of Lemma 2.2 can be satisfied for physically reasonable classes of relaxation functions β, γ. In this direction we have:

LEMMA 4.1. Let b_0, c_0, β, γ satisfy the assumptions of Lemma 2.1, Section 1. Define the kernel b in (V) by equation (b). Then the frequency domain assumption (F) of Lemma 2.2 is equivalent to the condition: there exists $\delta > 0$ such that

$$\operatorname*{Inf}_{(\eta\in R)} \frac{(c_0 - \mathrm{Re}\hat{\gamma}(i\eta))(b_0 + \mathrm{Re}\hat{\beta}(i\eta)) - \mathrm{Im}\hat{\gamma}(i\eta)\mathrm{Im}\hat{\beta}(i\eta)}{|b_0 + \hat{\beta}(i\eta)|^2} \geq \delta \tag{4.2}$$

Proof of Lemma 4.1. Define the constant $b_\infty > 0$ and the function B as in Lemma 2.1, Section 1. Taking the Laplace transform of B one computes

$$\hat{B}(i\eta) = -\frac{i}{\eta}\left(\frac{c_0 - \hat{\gamma}(i\eta)}{b_0 + \hat{\beta}(i\eta)} - b_\infty\right) \qquad (\eta \in R)$$

Thus

$$b_\infty - \eta \,\mathrm{Im}\, \hat{B}(i\eta) = \mathrm{Re}\left(\frac{c_0 - \hat{\gamma}(i\eta)}{b_0 + \hat{\beta}(i\eta)}\right) \qquad (\eta \in R)$$

from which the condition (4.2) is an immediate consequence.

Using Lemma 4.1 one can construct a large number of examples of functions β and γ such that assumption (F) is satisfied. In particular one has the following physically important special cases. Note that in Corollaries 4.2 and 4.3 below the physical conditions $b_0 + \int_0^t \beta(\tau)d\tau > 0$ $(0 \leq t < \infty)$, $c_0 - \int_0^t \gamma(\tau)d\tau > 0$ $(0 \leq t < \infty)$ are both satisfied (although they are not explicitly needed in the theory), because the functions β and γ are positive, and assumption (γ) is assumed to hold. For a different example in which (γ) is satisfied but the above physical conditions need not hold see Remark 4.8 below.

COROLLARY 4.2. Let $b_0 > 0$, $c_0 > 0$ and β, γ, $t\beta$, $t\gamma \in L^1(0,\infty)$. Also assume that β and γ are positive, nonincreasing and convex on $[0,\infty)$, and that the assumption (γ) is satisfied. Then assumption (F) is satisfied if either for a fixed $b_0 > 0$ the constant $c_0 > 0$ is chosen sufficiently large, or if for a fixed $c_0 > 0$ the constant $b_0 > 0$ is sufficiently large.

REMARK 4.3. (i) If $\beta = \gamma \equiv 0$ (the standard heat flow problem) (F) is satisfied for any choice of $b_0 > 0$, $c_0 > 0$ with $\delta = b_0/c_0$.

(ii) If $\beta \equiv 0$ and γ satisfies the assumptions of Corollary 4.2, (*F*) is satisfied for any choice of $b_0 > 0$, $c_0 > 0$ with $\delta = (c_0 - \int_0^\infty \gamma(t)dt)/b_0$.

(iii) If $\gamma \equiv 0$ and β satisfies the assumptions of Corollary 4.2, (*F*) is satisfied for any choice of $b_0 > 0$, $c_0 > 0$.

Sketch of Proof of Corollary 4.2. The proof will make use of Lemma 4.1; we establish (4.2). Since $\beta, \gamma \in L^1(0,\infty)$ and are positive, nonincreasing and convex, Re $\hat{\beta}(i\eta)$ and Re $\hat{\gamma}(i\eta)$ are nonnegative. The function

$$\mathrm{Im}\hat{\gamma}(i\eta)\,\mathrm{Im}\hat{\beta}(i\eta) = \int_0^\infty \gamma(t)\sin \eta t dt \int_0^\infty \beta(t)\sin \eta t dt \qquad (\eta \in R)$$

is even, continuous, zero when $\eta = 0$, nonnegative, and has limit zero as $\eta \to \infty$ (Riemann-Lebesgue lemma). The denominator in (4.2) satisfies

$$0 < b_0^2 \le |b_0 + \hat{\beta}(i\eta)|^2 \le 2b_0^2 + 3(\int_0^\infty \beta(t)dt)^2 \qquad (\eta \in R)$$

Moreover,

$$b_0 + \mathrm{Re}\hat{\beta}(i\eta) \ge b_0 > 0 \qquad (\eta \in R)$$

(so that (PW) is satisfied), and

$$c_0 - \mathrm{Re}\hat{\gamma}(i\eta) \ge c_0 - \int_0^\infty \gamma(t)dt > 0 \qquad (\eta \in R)$$

Therefore, the existence of $\delta > 0$ such that (4.2) holds is established for choices of b_0 and c_0 as asserted. This completes the proof.

Another physically important case for the heat flow problem is the following special case of Lemma 4.1 and Corollary 4.2.

COROLLARY 4.4. Let

$$\beta(t) = \sum_{k=1}^{n} b_k e^{-\beta_k t} \qquad (0 \le t < \infty)$$
$$\gamma(t) = \sum_{k=1}^{m} c_k e^{-\gamma_k t} \qquad (0 \le t < \infty) \tag{4.3}$$

with $b_k \ge 0$, $\beta_k \ge 0$, $c_k \ge 0$, $\gamma_k \ge 0$ and strict inequalities hold for at least one pair b_k, β_k and one pair c_k, γ_k. Let $b_0 > 0$, $c_0 > 0$, and $c_0 - \sum_{k=1}^{m} c_k/\gamma_k > 0$. Then the frequency domain condition (*F*) is satisfied if

$$b_0(c_0 - \sum_{k=1}^{m} \frac{c_k}{\gamma_k}) > \frac{5}{4}(\sum_{k=1}^{n} \frac{b_k}{\beta_k})(\sum_{k=1}^{m} \frac{c_k}{\gamma_k}) \tag{4.4}$$

The proof of Corollary 4.4 is a consequence of showing that there exists a $\delta > 0$ such that (4.2) holds. The inequality (4.4) follows by using elementary calculus to find the infimum over $\eta \in R$ of the expression in (4.2):

$$\frac{\left(c_0 - \sum_{k=1}^{m} \frac{c_k\gamma_k}{\gamma_k^2 + \eta^2}\right)\left(b_0 + \sum_{k=1}^{n} \frac{b_k\beta_k}{\beta_k^2 + \eta^2}\right) - \eta^2 \left(\sum_{k=1}^{m} \frac{c_k}{\gamma_k^2 + \eta^2}\right)\left(\sum_{k=1}^{n} \frac{b_k}{\beta_k^2 + \eta^2}\right)}{\left(b_0 + \sum_{k=1}^{n} \frac{b_k}{\beta_k^2 + \eta^2}\right)^2 + \left(\sum_{k=1}^{n} \frac{b_k\eta}{\beta_k^2 + \eta^2}\right)^2}$$

No claim is made that the constant 5/4 in (4.14) is optimal.

We next combine the properties stated in Lemmas 2.1, 2.2, and 2.3 of Section 1 and Lemma 4.1, and Corollaries 4.2 and 4.3 above with the abstract theory to establish the following result for the physical heat flow problem (4.1) in a one-dimensional material with memory. To see that a more general result (not necessarily physical) with β and γ oscillatory can hold we refer to Remark 4.8 below.

THEOREM 4.5. Let $b_0 > 0$, $c_0 > 0$, let $\beta, \gamma, t\beta, t\gamma \in L^1(0,\infty)$ and let $\beta \in L^2(0,\infty)$. Assume that β and γ are positive, nondecreasing, convex, and that

$$c_0 - \int_0^\infty \gamma(t)dt > 0 \qquad (\gamma)$$

Assume that $\sigma : R \to R$ satisfies assumptions (σ) of, Section 1, that the initial temperature $u_0 \in H_0^1(0,1)$, and that the external heat supply $h \in L^1(0,\infty;H) \cap L^2(0,\infty;H)$, where $H = L^2(0,1)$. Then the heat flow problem (4.1) has a unique strong solution u on $[0,\infty) \times (0,1)$ such that $\partial u/\partial t \in L^2_{loc}(0,\infty;H)$. Moreover, if either for a fixed $b_0 > 0$, the constant $c_0 > 0$ is sufficiently large, or for a fixed $c_0 > 0$, the constant $b_0 > 0$ is sufficiently large, then the solution u has the properties:

$$u \in L^\infty(0,\infty;H) \cap L^2(0,\infty;H), \ \frac{\partial u}{\partial t} \in L^2(0,\infty;H)$$

and $\lim_{t\to\infty} u(t) = 0$ strongly in H.

REMARK 4.6. For heat flow in more than one space dimension let Ω be a bounded body in R^2 or R^3 with smooth boundary Γ. Then the temperature u satisfies an equation of the form (4.1) with the operator $-\sigma(u_x)_x$ replaced by $-\nabla_0(\lambda(|\nabla u|)\nabla u)$; the boundary condition is $u(t,x) = 0$ $(0 \le t < \infty,\ x \in \Gamma)$, and the initial condition is $u(0,x) = u_0(x)$ $(x \in \Omega)$. If H is the Hilbert space $L^2(\Omega)$, if the function $\lambda : R \to R^+$ satisfies assumptions (λ) of, Section 1, if $u_0(x) \in H_0^1(\Omega)$, and $h \in L^1(0,\infty;H) \cap L^2(0,\infty;H)$, then the results of Theorem 4.5 holds, provided the constants $b_0 > 0$, $c_0 > 0$ and the relaxation functions β and γ satisfy the assumptions stated in Theorem 4.5.

REMARK 4.7. Let

$$\beta(t) = \sum_{k=1}^{n} b_k e^{-\beta_k t} \qquad (0 \le t < \infty)$$

$$\gamma(t) = \sum_{k=1}^{m} c_k e^{-\gamma_k t} \qquad (0 \le t < \infty)$$

where $b_k \ge 0$, $\beta_k \ge 0$, $c_k \ge 0$, $\gamma_k \ge 0$ and strict inequalities hold for at least one pair b_k, β_k and one pair c_k, γ_k. Let $b_0 > 0$, $c_0 > 0$, and $c_0 - \sum_{k=1}^{m} c_k/\gamma_k > 0$. Let σ, u_0, h satisfy the assumptions of Theorem 4.5. Then by Corollary 4.4 all conclusions of Theorem 4.5 hold if the inequality (4.4) is satisfied.

Proof of Theorem 4.5. Under the assumptions of the theorem the heat flow problem (4.1) is equivalent to the abstract Volterra equation (V) with the kernel b given by equation (b), the forcing term given by equation (f), and the operator $A = \partial\phi$ where $\phi : H \to (-\infty,\infty]$ is the proper, convex, l.s.c. function defined in Lemma 2.3, Section 1 (or Remark 2.4, Section 1 in more than one space dimension). To establish the existence and uniqueness of a strong solution of (V) (equivalent to (4.3)), we apply Theorem 3.10, Section 2. Lemma 2.2, Section 1, shows that the assumptions of Theorem 3.10, Section 2 concerning f are satisfied with $f(0,x) = u_0(x) \in H_0^1(0,1) = D(\phi)$. Moreover, (H_ϕ) is satisfied. Lemma 2.1, Section 1, shows that assumptions (H_b) are satisfied. Thus to apply Theorem 3.10, Section 2, we must still verify that $B' \in BV_{loc}[0,\infty)$. From the expression for B in Lemma 2.1, Section 1, we compute

$$B'(t) = -\frac{\gamma(t)}{b_0} + c_0\rho(t) - (\gamma*\rho)(t) \qquad (0 \le t < \infty) \tag{4.5}$$

Since β is monotone by hypothesis, the resolvent equation (ρ) and a standard argument (see e.g. Bellman and Cooke [12]) show that ρ is monotone. Finally, since γ is monotone, it follows that $B' \in BV[0,\infty)$. Thus Theorem 3.10, Section 2, yields the existence and uniqueness of a strong solution u of (V) on $[0,\infty)$ such that $u' \in L^2_{loc}[0,\infty;H)$.

We shall next apply Theorem 3.1 of this section. Concerning the kernel b, Lemma 2.1, Section 1, shows that assumptions (H_b) are satisfied with

$b_\infty > 0$. Moreover, Corollary 4.2 shows that b satisfies the frequency domain condition (F) if b_0 and c_0 are chosen as in the statement of Theorem 4.5.

Lemma 2.3, Section 1 (or Remark 2.4, Section 1 in the case of more than one dimension) shows that $\phi(y) \geq 0$ $(y \in H)$, $\lim_{|y|\to\infty} \phi(y) = +\infty$, and that the inclusion $\partial\phi(w) \ni 0$ has $w = 0$ as the only solution. Lemma 2.2, Section 1, shows that assumptions (H_f) are satisfied. Therefore, by Theorem 3.1 the solution u has the properties:

$$\sup_{0\leq t<\infty} \phi(u(t)) < \infty, \qquad \sup_{0\leq t<\infty} |u(t)| < \infty, \qquad \frac{du}{dt} \in L^2(0,\infty), \qquad u(t)$$

is uniformly continuous on $[0,\infty)$, $\lim_{t\to\infty} \phi(u(t)) = 0$,

and $u(t) \to 0$ (weakly) as $t \to \infty$

Lemma 2.3, Section 1 also shows that under assumption (σ) the coercivity assumption (2.4) is satisfied for every $T > 0$ with $\varepsilon = p_0\pi^2 > 0$ (or another positive constant in the case of more than one space dimension--see Remark 2.4, Section 1). Therefore, by Theorem 3.2 one also has $u \in L^2(0,\infty;H)$ and $u(t) \to 0$ as $t \to \infty$ strongly in H. This completes the proof.

REMARK 4.8. Suppose

$$\beta(t) = b_1 e^{-\beta_1 t}\cos\lambda t \qquad (b_1, \beta_1 > 0,\ 0 \leq t < \infty)$$

$$\gamma(t) = c_1 e^{-\gamma_1 t}\cos\omega t \qquad (c_1, \gamma_1 > 0,\ 0 \leq t < \infty)$$

and assume that $b_0 > 0$, $c_0 > 0$. Also suppose that σ and h satisfy the assumptions of Theorem 4.5. Although the assumptions concerning β, γ in Theorem 4.5 are not satisfied, one still has by Lemma 2.1, Section 1, that assumptions (H_b) hold with $b_\infty > 0$ provided

$$c_0 - \frac{c_1\gamma_1}{\gamma_1^2 + \omega^2} > 0 \tag{γ}$$

Moreover, $B' \in BV_{loc}[0,\infty)$ from (4.5), and the existence and uniqueness of a strong solution of (V) (equivalent to (4.1)) such that $u' \in L^2_{loc}(0,\infty;H)$ follows from Theorem 3.10, Section 2. Thus to obtain all of the conclusions of Theorem 4.5 we need only verify that the frequency domain assumption (F) holds in order to apply Theorems 3.1 and 3.2. By Lemma 4.1 it suffices to find sufficient conditions on the constants b_0, c_0, b_1, β_1, λ, c_1, γ_1, ω so that (4.2) holds. An elementary, but tedious calculation shows that $\delta > 0$ such that (4.2) holds exists in the case $\gamma_1^2 > \omega^2$, $\beta_1^2 > \lambda^2$, provided assumption (λ) above holds, and provided $b_0 > 0$ is chosen sufficiently large.

While no claim is made here that the above functions β and γ represent physically plausible relaxation functions, it is of some interest that the theory can still be applied. In this connection it may also be noted that here the function

$$C(t) = c_0 - \int_0^t \gamma(\tau)d\tau = c_0 - \frac{c_1\gamma_1}{\gamma_1^2 + \omega^2}(1 - e^{-\gamma_1 t}\cos\omega t) - \frac{c_1\omega}{\gamma_1^2 + \omega^2} e^{-\gamma_1 t}\sin\omega t$$

In a genuinely physical problem as motivated above one would need to require $C(t) > 0$ $(0 \le t < \infty)$, as well as assumption (γ). However, in the application of the theory the physical requirement $C(t) > 0$ $(0 \le t < \infty)$ is not used and indeed, for example,

$$C(\tfrac{\pi}{2}) = c_0 - \frac{c_1\gamma_1}{\gamma_1^2 + \frac{\pi^2}{4}} - \frac{\frac{\pi}{2}c_1}{\gamma_1^2 + \frac{\pi^2}{4}} e^{-\gamma_1 \frac{\pi}{2}}$$

could be negative, even though $C(\infty) = c_0 - c_1\gamma_1/(\gamma_1^2 + \pi^2/4) > 0$ holds.

4. EXISTENCE AND ASYMPTOTIC BEHAVIOR OF POSITIVE SOLUTIONS OF A NONLINEAR VOLTERRA EQUATION FOR HEAT FLOW

4.1. Introduction

In Section 4 we discuss the positivity of solutions and their asymptotic behavior as $t \to \infty$, of the nonlinear Volterra equation

$$u(t) + (b*Au)(t) \ni f(t) \qquad (0 \le t < \infty) \tag{V}$$

in the general setting: $b : [0,\infty) \to R$ is a given kernel, A is a nonlinear (possibly multivalued) m-accretive operator defined on a real Banach space X, $f : [0,\infty) \to X$ is a given function; the integral in (V) is understood in the sense of Bochner. The assumptions which are imposed on b, A, f are motivated by the problem of nonlinear heat flow in a material with memory formulated in Section 1 to which the general positivity and asymptotic theory developed in Section 4.3 will be applied in Section 4.4. A different application of general theory is given in Example 3.4, Section 4.3, to a nonlinear conservation law with memory.

Section 4, which generalizes and complements earlier work of Clément and Nohel [18] on positivity and of Clément [17] on limiting behavior of positive solutions of (V), is primarily based on a forthcoming paper by Clément and Nohel [19]. The generalization enables us to apply the theory to the physical problem in Section 4.4. General existence, uniqueness and continuous dependence results for solutions of (V) which need not be positive have been established by Crandall and Nohel [26] and by Gripenberg [34] (see also Theorems 3.5 and 3.7, Section 3); these will be referred to as needed.

We will motivate the assumptions on the kernel b which are needed for positivity of solutions of (V) and which will be needed throughout the analysis by means of a simple linear problem at the end of Section 4.1. These considerations suggest the concept of complete positivity of the kernel b (Definition 1.1 below) which plays an important role in the analysis. Some properties and a useful characterization of completely positive kernels are obtained in Section 4.2.

We shall consider equation (V) in the slightly less general form

$$u(t) + (b*Au)(t) \ni u_0 + (b*g)(t) \qquad (0 \le t < \infty) \tag{V_g}$$

We assume throughout the following minimal assumptions:

$$\begin{cases} b \in L^1_{loc}(0,\infty), \\ A \text{ m-accretive in a real Banach space } X, \\ u_0 \in \overline{D(A)}, \qquad \text{and} \qquad g \in L^1_{loc}(0,\infty;X) \end{cases} \tag{H_1}$$

The motivation for taking $f = u_0 + b*g$ in (V) is given in Section 4.3 (see the argument at the beginning of Section 4.3 following (V_g)). The main results of Section 4, described in Section 4.3, give a rather complete description of the asymptotic behavior of the positive solutions of the abstract equation (V) as $t \to \infty$, including a priori estimates for their rates of decay. The results are then applied to the physical problem in Section 4.4.

The additional assumption we shall make on the kernel b in order to insure positivity of solutions was first introduced in [18]; it is motivated by the following remark. If $b \equiv 1$ then (V_g) reduces to the evolution equation

$$\begin{cases} \frac{du}{dt} + Au \ni g \\ u(0) = u_0 \end{cases} \tag{DE}$$

It is well-known [13] that if the resolvent $J_\lambda = (I + \lambda A)^{-1}$ of A maps a closed convex cone P of X into itself for every $\lambda > 0$, then $u(t) \in P$ for all $t > 0$, provided that $u_0 \in P$ and $g(t) \in P$ a.e. on $[0,\infty)$. Let us take for instance

$$X = \{u \in C[a,b] | u(a) = u(b) = 0\}$$

equipped with the supremum norm; $D(A) = \{u \in X \mid u \in C^2[a,b] \text{ and } u_{xx} \in X\}$ and $Au(x) = -u_{xx}(x)$ for $u \in D(A)$. It is standard that A is m-accretive in X. Moreover, if $P = \{u \in X | u(x) \geq 0\ x \in [a,b]\}$, then $J_\lambda P \subset P$ for every $\lambda > 0$; thus, as is classical, the solution of the heat equation is nonnegative provided that the initial value u_0 and the forcing form g are nonnegative.

We want to consider a class of kernels b under which the solution (V), resp. (V_g), preserves this positivity property. This requirement is useful and natural in the application to the model of heat flow in a material with memory discussed in Section 4.4 and in Example 3.4 of Section 4.3.

Consider (V_g) with $Au = -u_{xx}$ with D(A) as in the above example. It is easy to give necessary conditions to be imposed on b in order that positivity is preserved by (V_g) whenever u_0 and g are positive. Let $\bar{\lambda}$ denote the principal eigenvalue and $\bar{u}$ the corresponding principal eigenfunction of A, normalized by $\max_{x\in[a,b]} \bar{u}(x) = 1$. Clearly $\bar{\lambda} = (\pi/b-a)^2$ and $\bar{u}(x) = \sin(\pi/b-a)(x-a)$. If $u_0 = \alpha\bar{u}$, $g(t) = \bar{\lambda}\beta(t)\bar{u}$ with $\alpha > 0$, and $\beta(t) \geq 0$ where $\beta \in L^1_{loc}(0,\infty)$, then, as can be verified directly, the strong solution of (V_g) is

$$u(t) = [\alpha s(\bar{\lambda}b)(t) + (\beta * r(\bar{\lambda}b))(t)]\bar{u} \qquad (0 \leq t < \infty) \tag{1.1}$$

where the functions s(b) and $r(b):[0,\infty) \to R$ are respectively solutions of the linear Volterra equations

$$s(b)(t) + (b*s(b))(t) = 1 \qquad (0 \leq t < \infty) \tag{s(b)}$$

$$r(b)(t) + (b*r(b))(t) = b(t) \qquad (0 \leq t < \infty) \tag{r(b)}$$

Recall the standard fact (see e.g. R. K. Miller [63]) that if $b \in L^1_{loc}(0,\infty)$, the functions s(b), r(b) are uniquely defined and s(b), $r(b) \in L^1_{loc}(0,\infty)$. Moreover, if $F \in L^1_{loc}(0,\infty)$ the unique solution of the linear Volterra equation

$$u(t) + (b*u)(t) = F(t) \qquad (0 \leq t < \infty) \tag{1.2}$$

is given by

$$u(t) = F(t) - (r(b)*F)(t) \qquad (0 \le t < \infty) \tag{1.3}$$

In particular, taking $F \equiv 1$ in (1.2), one has

$$s(b)(t) = 1 - \int_0^t r(b)(\tau)d\tau \qquad (0 \le t < \infty) \tag{1.4}$$

so that $s(b)$ is absolutely continuous on $[0,\infty)$ whenever $b \in L^1_{loc}(0,\infty)$. The function $s(b)$ is called the fundamental solution of (1.2), while the function $r(b)$ is called the resolvent kernel associated with b.

Returning to the solution (1.1) of (V_g) with $Au = -u_{xx}$, $\overline{\lambda} > 0$, $\overline{u}$, u_0, g defined above, we note that $\overline{u}(x) > 0$ for $x \in (a,b)$. Thus the solution u(t) will be nonnegative for every $\alpha \ge 0$ and for every $\beta \in L^1_{loc}(0,\infty)$, $\beta \ge 0$, only if the functions $r(b)$, $s(b)$ are nonnegative on $[0,\infty)$. Moreover, if one imposes the requirement that the solution (1.1) of (V_g) should be nonnegative and independent of the length of the interval (a,b), it is clear that both of the functions $r(\lambda b)$ and $s(\lambda b)$ must be nonnegative for every $\lambda > 0$. We remark that these latter necessary conditions imposed on the kernel b have been shown to be sufficient to guarantee the preservation of positivity by the solution operator of the nonlinear equation (V) in the general case of A m-accretive on X (see [18, Theorem 4.5]).

The above considerations suggest the following concept of complete positivity of the kernel b:

DEFINITION 1.1. We shall say that the kernel b is completely positive on $[0,T]$ if $b \in L^1(0,T)$ and if the functions $r(\lambda b)$ and $s(\lambda b) = 1 - 1*r(\lambda b)$ are nonnegative on $[0,T]$ for every $\lambda > 0$.

Some known sufficient conditions which insure the complete positivity of the kernel on $[0,T]$ are:

(i) $b \in L^1(0,T)$ is nonnegative, nonincreasing, and log b is convex (see Miller [62], Levin [52], Clément and Nohel [18]).

(ii) (special case of (i)) $b \in L^1(0,T)$ and b is completely monotonic on $(0,T)$ (see Miller [62]).

4.2. Completely Positive Kernels

In this section we give an alternate and useful characterization of completely positive kernels (Theorem 2.2) which will be needed for the development of the asymptotic properties of positive solutions of the abstract Volterra equation (V_g). For this purpose we consider the linear scalar Volterra equation (1.2) in the form

$$u + b*u = u_0 + b*g \tag{2.1}$$

where $b \in L^1(0,T)$, $u_0 \in R$, $g \in L^1(0,T)$, and $T > 0$. Its unique solution (see (1.3) and (1.4)) is given by

$$u(t) = u_0 s(b)(t) + (r(b)*g)(t) \qquad (0 \leq t \leq T) \tag{2.2}$$

In the following proposition we list some elementary properties of completely positive kernels which are needed in the sequel.

PROPOSITION 2.1. Assume that b is completely positive on $[0,T]$ for some $T > 0$. Then:

(1) b is nonnegative on $[0,T]$ and for every $\mu > 0$, $s(\mu b)$ is nonnegative and nonincreasing on $[0,T]$.

(2) For every $\mu > 0$, $r(\mu b)$ is itself completely positive on $[0,T]$.

Next, assume b is completely positive on $[0,T]$ for every $T > 0$. Then:

(3) If $b \in L^1[0,\infty)$, then for every $\mu > 0$

$$\lim_{t\to\infty} s(\mu b)(t) = \left(1 + \mu \int_0^\infty b(\tau)d\tau\right)^{-1}$$

$$\| r(\mu b) \|_{L^1(0,\infty)} = \mu \int_0^\infty b(\tau)d\tau \left(1 + \int_0^\infty b(\tau)d\tau\right)^{-1}$$

(4) If $b \notin L^1(0,\infty)$, then for every $\mu > 0$

$$\lim_{t\to\infty} s(\mu b)(t) = 0 \quad \text{and} \quad \| r(\mu b) \|_{L^1(0,\infty)} = 1$$

(5) If $b \notin L^1(0,\infty)$ and $b \in AC[0,\infty)$, then for every $\mu > 0$, $r(\mu b) \in C[0,\infty)$ and $\lim_{t\to\infty} r(\mu b)(t) = 0$.

The proof of Proposition 2.1 is elementary and is omitted; for details see [19].

In the next result we give an alternate and useful characterization of completely positive kernels b. Some arguments used are similar to those of [33].

THEOREM 2.2. Let $T > 0$, $b \in L^1(0,T)$, $b \not\equiv 0$. Then b is completely positive on $[0,T]$ if and only if there exists $\alpha \geq 0$ and $k \in L^1(0,T)$ nonnegative and nonincreasing satisfying:

$$\alpha b(t) + k*b(t) = 1 \qquad t \in [0,T] \tag{2.3}$$

REMARKS. (i) It follows from (2.3) that $\alpha > 0$ if and only if $b \in L^\infty(0,T)$. If this is the case $b = \alpha^{-1}s(\alpha^{-1}k)$ and thus $b \in AC[0,T]$. Conversely if $b \in AC[0,T]$, then $\alpha = b(0)^{-1} > 0$. Moreover, observe that if $\alpha > 0$, then $k \in BV[0,T]$ (equivalently $k(0^+) < \infty$) if and only if $b' \in BV[0,T]$).

The importance of the remark $\alpha > 0$ ($k \in BV[0,T]$) is that for kernels b satisfying the assumption:

$$b \in AC[0,T], \qquad b(0) > 0, \qquad b' \in BV[0,T] \tag{H}$$

the existence and uniqueness of a generalized solution $u \in C([0,T]; \overline{D(A)})$ of the abstract Volterra equation (V) has been established by Crandall and Nohel (Section 2, Theorem 3.5), whenever the operator A is m-accretive, and $f(0) \in \overline{D(A)}$, $f \in W^{1,1}(0,T;X)$. For the special case $X = H$ a real Hilbert space and $A = \partial\phi$ we refer to Section 2, Theorem 3.10. Recently Gripenberg [34, Theorem 2] has extended this result to the case of kernels $b = b_1 + b_2$, where b_1 satisfies the above regularity assumption and where $b_2 \in L^1(0,T)$, b_2 is positive, nonincreasing, and log b_2 is convex on $(0,T)$ with A and f as above (see Section 2, Theorem 3.7). This result with $b_1 \equiv 0$ and A linear was already established by Clément and Nohel [18]. These more general completely positive kernels b correspond to the case $\alpha = 0$. The problem of existence of generalized solutions of (V) with only the assumption that b is completely positive is under study and will be treated elsewhere.

(ii) It follows from Theorem 2.2 and Remark (i) that if b is completely positive, then b need not be nonincreasing; it also need not be convex and a fortiori log convex. Choose $\alpha = 1$ and $k(t) = 1$ for $t \in [0,1]$ and $k(t) = 0$ for $t > 1$; then $b = s(k)$ is completely positive. But as shown in Levin [52], $b' = -r(k)$ is negative on some interval $[0,\alpha)$ with $\alpha \in (1,2)$ and positive in $(\alpha,2]$. Thus b is not nonincreasing on [0,2]. Moreover, assume b to be convex on $[0,\infty)$. Then b is strictly increasing for $t > \alpha$, and moreover $\lim_{t\to\infty} b(t) = \infty$. But this is impossible, since $b(t) \leq 1$ as seen from (2.3) and the fact that k,b are nonnegative and $\alpha = 1$. Thus b is not convex.

(iii) If b is completely positive and absolutely continuous on [0,T], then it follows from (2.3) that $b(t) \leq b(0)$ for $t \in [0,T]$.

(iv) It follows from Theorem 2.2 that if $b \in L^1_{loc}(0,\infty)$, b is positive, decreasing, log b is convex and $b(0^+) = \infty$, then the linear Volterra equation of the first kind

$$k*b(t) = 1 \quad t > 0 \tag{2.4}$$

possesses a unique solution $k \in L^1_{loc}(0,\infty)$ which is nonnegative and nonincreasing. However, given $k \in L^1[0,T]$, k nonnegative and nonincreasing, equation (2.4) may not have a solution in $L^1(0,T)$. (Take $k(t) \equiv 1$). Thus when $\alpha = 0$, equation (2.3) does not provide a way to generate completely positive kernels which are not absolutely continuous on [0,T].

We omit the quite technical proof of Theorem 2.2. It makes repeated use of the following result due to Levin [52]. If u satisfies $u + b*u = f$ with $b \in L^1(0,T)$, b nonnegative and nonincreasing, $f \in L^1(0,T)$, nonnegative, nondecreasing, then u is nonnegative on [0,T]. The proof of Theorem 2.2 may be found in [19].

4.3. Qualitative Properties of Abstract Volterra Equations with Completely Positive Kernels

In this section we study some properties of generalized solutions, including positivity and the asymptotic behavior of positive solutions as $t \to \infty$, of the nonlinear abstract Volterra equation

$$u + b*Au \ni u_0 + b*g \qquad (t \geq 0) \tag{V_g}$$

Although our results are stated for generalized solutions, it is obvious that the results hold for strong solutions whenever strong solutions are shown to be generalized solutions (see Remark (i) following Theorem 3.1 below).

The justification for taking $f = u_0 + b*g$ in (V) is as follows. If b satisfies assumption (H) (Section 4.2, Remark (i) following Theorem 2.2), if $f \in W^{1,1}(0,T;X)$, and if $f(0) \in \overline{D(A)}$, then there exists a unique $u_0 \in \overline{D(A)}$ and a unique $g \in L^1(0,T;X)$ such that

$$f(t) = u_0 + (b*g)(t) \qquad (0 \leq t \leq T) \tag{3.1}$$

Indeed, $u_0 = f(0) \in \overline{D(A)}$ and g is the unique solution of the linear equation

$$b(0)g(t) + (b'*g)(t) = f'(t) \qquad (0 \leq t \leq T) \tag{3.2}$$

Conversely, if b satisfies assumption (H) and $u_0 \in \overline{D(A)}$, $g \in L^1(0,T;X)$, then f given by (3.1) satisfies $f(0) \in \overline{D(A)}$, $f \in W^{1,1}(0,T;X)$. We shall make the following general assumptions:

$$\left\{\begin{array}{l} A \text{ is m-accretive in } X \\ u_0 \in \overline{D(A)} \\ g \in L^1_{loc}(0,\infty;X) \\ b \text{ is completely positive on } [0,\infty) \end{array}\right. \qquad (\tilde{H})$$

The basic preliminary result assuming the global existence of solutions of (V_g) under assumption $(\tilde{H})$ is known:

THEOREM 3.1. If A, u_0, g and b satisfy assumption $(\tilde{H})$ then:

(1) if u_1 and u_2 are the generalized solutions of (V_g) corresponding to the data $u_{0,i}$, g_i, $i = 1,2$, then the following estimate holds:

$$\| u_1(t) - u_2(t) \| \leq \| u_{0,1} - u_{0,2} \| + (b^* \|g_1 - g_2\|)(t) \qquad t \geq 0 \quad \text{a.e.} \tag{3.3}$$

(2) if P is a closed convex cone in X, if $J_\lambda(P) \subset P$ for every $\lambda > 0$, and if $u_0 \in P$ and $g(t) \in P$ a.e. on $[0,\infty)$, then $u(t) \in P$ a.e. on $[0,\infty)$; moreover, if $v - u \in P$ implies $J_\lambda v - J_\lambda u \in P$ for every $\lambda > 0$, $u,v \in X$, and if $u_{0,2} - u_{0,1} \in P$, $g_2(t) - g_1(t) \in P$ a.e. on $[0,\infty)$, then u_i, $i = 1,2$, the corresponding generalized solutions of (V_g) satisfy $u_2(t) - u_1(t) \in P$, a.e. on $[0,\infty)$.

REMARKS. (i) The existence of a generalized solution in the linear case under the assumption b completely positive was proved in [18]. In the nonlinear case, when $b \in AC[0,T]$, $b(0) > 0$ and $b \in BV[0,T]$, or when $b \in L^1(0,T)$, b is positive, nonincreasing and log b is convex on $(0,T)$, the existence of generalized solutions of (V_g) follows from results of Crandall and Nohel [26] and Gripenberg [34] already discussed in Section 2 and in Remark (i) following Theorem 2.2. Moreover, if more regularity is assumed on b and f, then (see [26,34]) the generalized solution is also a strong solution of (V_g).

(ii) Estimate (3.3) was proved in [17].

(iii) We sketch the proof of the positivity result asserted in (2); the details appear in Clément and Nohel [18]; the last assertion in (2) can be established in the same way.

Consider the approximating equation of (V_g) resulting from replacing the operator A by its Yosida approximation

$$A_\lambda = \frac{1}{\lambda}(I - J_\lambda), \qquad J_\lambda = (I - \lambda A)^{-1}:$$

$$u_\lambda + b*A_\lambda u_\lambda = u_0 + b*g \qquad (0 \le t < \infty) \tag{V_λ}$$

By the definition of A_λ equation (V_λ) is the same as

$$u_\lambda + \frac{1}{\lambda} b*u_\lambda = u_0 + b*g + \frac{1}{\lambda} b*J_\lambda u_\lambda$$

Using (2.2) with g replaced by $g + 1/\lambda\, J_\lambda u_\lambda$, one easily checks that (V_λ) is equivalent to the integral equation

$$u_\lambda = f_\lambda + W_\lambda(u_\lambda) \tag{$\overline{V}_\lambda$}$$

where

$$f_\lambda = s(\frac{1}{\lambda} b)u_0 + \lambda r(\frac{1}{\lambda} b)*g$$

and

$$W_\lambda(u_\lambda)(t) = r(\frac{1}{\lambda} b)*J_\lambda(u_\lambda)$$

With

$$v_\lambda = u_\lambda - f_\lambda$$

the integral equation $(\overline{V}_\lambda)$ is equivalent to the nonlinear equation

$$v_\lambda = W_\lambda(v_\lambda + f_\lambda) \tag{$\overline{\overline{V}}_\lambda$}$$

Let $0 < T < \infty$ be arbitrary. By the complete positivity of b and by the hypothesis of Theorem 3.1, part (2), $f_\lambda(t) \in P$ a.e. on $[0,T]$. Noting that W_λ maps $L^1(0,T;X)$ into itself and recalling that the operator J_λ is a contraction (since A is m-accretive), one proves (for details see [18] and [26]) that some iterate W_λ^n of W_λ is a strict contraction on $L^1(0,T;X)$ for n sufficiently large. Thus the integral equation $(\overline{\overline{V}}_\lambda)$ has a unique solution $v_\lambda \in L^1(0,T;X)$ given in

$$v_\lambda = \lim_{n\to\infty} W_\lambda^n(v_0) \qquad \text{for any} \qquad v_0 \in L^1(0,T;X)$$

Therefore, the approximating equation (V_λ) has the unique solution $u_\lambda = f_\lambda + v_\lambda$. But $f_\lambda \in P$ a.e. on $[0,T]$, and if $v_0(t) \in P$ a.e. on $[0,T]$, the complete positivity of b and the assumption $J_\lambda(P) \subset P$ for every $\lambda > 0$ insure that $W_\lambda(v_0)(t) \in P$ a.e. on $[0,T]$ and the same holds for $W_\lambda^n(v_0)(t)$ for every n. Consequently $u_\lambda(t) \in P$ a.e. on $[0,T]$, and if u is a solution of (V_g) on $[0,T]$ such that $u = \text{weak} \lim_{\lambda\to 0} u_\lambda$ in $L^1(0,T;X)$, then $u(t) \in P$ a.e. on $[0,T]$. Since $T > 0$ is arbitrary this completes a sketch of the proof.

We next obtain some results concerning the asymptotic behavior of solutions of (V_g) as $t \to \infty$. We first consider the case $b \in L^1(0,\infty)$.

THEOREM 3.2. Let A, u_0, g, b, satisfy the general assumptions $(\tilde{H})$ with $b \not\equiv 0$ and $b \in L^1(0,\infty)$.

(1) Let $g \in L^\infty(0,\infty;X)$ and assume there exists $g^\infty \in X$ such that $\lim_{t\to\infty} \| g(t) - g^\infty \| = 0$. Let u be the generalized solution of (V_g) and define $u^\infty = J_{\bar{b}}(u_0 + \bar{b}g^\infty)$, where $\bar{b} = \int_0^\infty b(t)dt > 0$. Then the following estimate, which implies strong convergence of u(t) to u^∞ as $t \to \infty$, holds:

$$\| u(t) - u^\infty \| \leq \frac{\int_t^\infty b(\tau)d\tau}{\bar{b}} \| u_0 - u^\infty \| + (b* \| g - g^\infty \|)(t) \tag{3.4}$$

$$(0 \leq t < \infty)$$

(2) In addition, let $b \in L^\infty(0,\infty)$ and $\lim_{t\to\infty} b(t) = 0$. Let $g = g_1 + g_2$, where g_1 satisfies the assumptions of g in part 1, and where $g_2 \in L^1(0,\infty;X) \cap L^p(0,\infty;X)$, $P \in (1,\infty)$. Let u be the generalized solution of (V_g) and let $u^\infty = J_{\bar{b}}(u_0 + \bar{b}g_1^\infty)$. Then the following estimate, which implies strong convergence of u(t) to u^∞ as $t \to \infty$, holds:

$$\| u(t) - u^\infty \| \leq \frac{\int_t^\infty b(s)ds}{\bar{b}} \| u_0 - u^\infty \| + (b* \| g_1 - g_1^\infty \|)(t) + (b* \|g_2\|)(t) \qquad (0 \leq t < \infty) \tag{3.5}$$

where $g_1^\infty = \lim_{t\to\infty} g_1(t)$.

REMARK. Part 2 is proved below. Part 1 of Theorem 3.2 was proved in Clément [17] and the proof will be omitted; it uses ideas similar to the proof of part 2.

Next we consider the case where $b \notin L^1(0,\infty)$ which is needed for the application in Section 4.4. In order to establish the strong convergence of u to u^∞ as $t \to \infty$ we shall require that the nonlinear operator A in (V) satisfy a rather strong coercivity condition.

THEOREM 3.3. Let A, u_0, g and b satisfy the general assumptions $(\tilde{H})$ with $b \notin L^1(0,\infty)$ and suppose A is coercive in the sense that there exists $\omega > 0$ for which $A - \omega I$ is accretive in X.

(1) Let g be in $L^\infty(0,\infty;X)$ and let $g^\infty \in X$ such that $\lim_{t\to\infty} \| g(t) - g^\infty \| = 0$. Let u be the generalized solution of (V_g) and let u^∞ be the unique element in X satisfying $Au^\infty \ni g^\infty$. Then the following estimate which implies strong convergence of u(t) to u^∞ as $t \to \infty$, holds;

$$\begin{aligned} \| u(t) - u^\infty \| &\leq \textstyle\int_t^\infty r(\omega b)(\tau)d\tau \; \| u_0 - u^\infty \| \\ &\quad + \omega^{-1}(r(\omega b) * \| g - g^\infty \|)(t) \qquad (0 \leq t < \infty) \end{aligned} \tag{3.6}$$

(2) In addition, let b be AC$[0,\infty]$ and $g = g_1 + g_2$ where g_1, g_2 satisfy the assumptions of Theorem 3.2, part (2), with $g_1^\infty = \lim_{t\to\infty} g_1(t)$. Let u be the generalized solution of (V_g), and let u^∞ be the unique element in X satisfying $Au^\infty \ni g_1^\infty$. Then the following estimate, which implies strong convergence of u(t) to u^∞ as $t \to \infty$, holds;

$$\begin{aligned} \| u(t) - u^\infty \| &\leq \textstyle\int_t^\infty r(\omega b)(\tau)d\tau \; \| u_0 - u^\infty \| \\ &\quad + \omega^{-1}(r(\omega b) * \| g_1 - g_1^\infty \|)(t) \\ &\quad + \omega^{-1}(r(\omega b) * \| g_2 \|)(t) \end{aligned} \tag{3.7}$$

REMARKS. (i) Since b is completely positive and $b \notin L^1(0,\infty)$, it follows (see Proposition 2.1) that $r\,(\omega b) \in L^1(0,\infty)$, and therefore, if the assumptions of part 1 hold, (3.6) implies $\lim_{t\to\infty} \| u(t) - u^\infty \| = 0$.

When b also satisfies $b \in AC[0,\infty]$ it follows [see Proposition 2.1] that $r(\omega b) \in L^1(0,\infty) \cap C[0,\infty)$, and $\lim_{t\to\infty} r(\omega b)(t) = 0$. Therefore (3.7) implies $\lim_{t\to\infty} \| u(t) - u^\infty \| = 0$, if the assumptions of part 2 hold.

(ii) As is clear from the proofs, the assumption $g \in L^\infty(0,\infty;X)$ and there exists g^∞ such that $\lim_{t\to\infty} \| g(t) - g^\infty \| = 0$ in part (1) of Theorem 3.2 can be weakened to $g \in L^1_{loc}(0,\infty;X)$ and there exists $g^\infty \in X$ such that $\lim_{t\to\infty} (b * \| g - g^\infty \|)(t) = 0$. Similar generalizations can be made in Theorem 3.2, part (2) and in Theorem 3.3.

Proof of Theorem 3.2, part 2. As in the proof of Theorem 3.2, part (1) in [17], we first prove the result with A replaced by A_λ, $\lambda > 0$, and then we pass to the limit as $\lambda \downarrow 0$. For $\lambda > 0$, let u_λ be the strong solution of the approximating equation

$$u_\lambda + b*A_\lambda u_\lambda = u_0 + b*g \qquad (t \in [0,\infty))$$

Using the definition of A_λ and applying (2.2) we see that u_λ satisfies the equation

$$u_\lambda = r(\lambda^{-1}b)*J_\lambda u_\lambda + s(\lambda^{-1}b)u_0 + r(\lambda^{-1}b)*g \tag{3.8}$$

for $t \in [0,\infty)$. Since A_λ is also m-accretive, there is a unique u_λ^∞ satisfying the limiting equation

$$u_\lambda^\infty + \overline{b}A_\lambda u_\lambda^\infty = u_0 + \overline{b}g_1^\infty \tag{3.9}$$

Using the fact that $b \in L^1(0,\infty)$ and $g = g_1 + g_2$ we can rewrite (3.9) in the equivalent form

$$u_\lambda^\infty + b*A_\lambda u_\lambda^\infty = u_0 + b*g + b*(g_1^\infty - g_1) - b*g_2 - \xi w_\lambda^\infty \tag{3.10}$$

where

$$\xi(t) = \int_t^\infty b(s)ds \qquad \text{and} \qquad w_\lambda^\infty = A_\lambda u_\lambda^\infty - g_1^\infty$$

Let $\eta : [0,\infty) \to R$ be the unique solution of the linear equation

$$\eta + \lambda^{-1}b*\eta = \xi$$

then obviously

$$\eta w_\lambda^\infty + \lambda^{-1}b*\eta w_\lambda^\infty = \xi w_\lambda^\infty \qquad (0 \le t < \infty) \tag{3.11}$$

Using (3.10), (3.11), (2.2) and the definition of A_λ we obtain

$$\begin{aligned} u_\lambda^\infty = {} & r(\lambda^{-1}b)*J_\lambda u_\lambda^\infty + s(\lambda^{-1}b)u_0 + \lambda r(\lambda^{-1}b)*g \\ & + \lambda r(\lambda^{-1}b)*(g_1^\infty - g_1) + \lambda r(\lambda^{-1}b)*g_2 - \eta w_\lambda^\infty \end{aligned} \tag{3.12}$$

Subtracting (3.12) from (3.8) we obtain

$$\begin{aligned} \| u_\lambda(t) - u_\lambda^\infty \| \le {} & (r(\lambda^{-1}b)*\| u_\lambda^\infty - u_\lambda \|)(t) \\ & + \lambda(r(\lambda^{-1}b)*\| g_1^\infty - g_1 \|)(t) + \lambda(r(\lambda^{-1}b)*\| g_2 \|(t) \\ & + |\eta|(t) \; \| w_\lambda^\infty \| \end{aligned} \tag{3.13}$$

It is shown in Clément [17] (see argument following (3.18) in [17]) that $\eta \geq 0$. Thus by using the same argument as in [17] one gets (take convolution of (3.13) with $\lambda^{-1}b$):

$$\begin{aligned} \| u_\lambda(t) - u_\lambda^\infty \| &\leq \xi(t) \| w_\lambda^\infty \| + (b* \| g_1 - g_1^\infty \|)(t) \\ &+ (b* \| g_2 \|)(t) \qquad (0 \leq t < \infty) \end{aligned} \tag{3.14}$$

The conclusion (3.5) follows by using (3.9) and rewriting

$$\xi(t) \| w_\lambda^\infty \| = \frac{\int_t^\infty b(s)ds}{\int_0^\infty b(s)ds} \| u_0 - u_\lambda^\infty \|$$

and then letting $\lambda \downarrow 0$. Note that

$$u^\infty = (I + \bar{b}A)^{-1}(u_0 + \bar{b}g_1^\infty) = \lim_{\lambda \downarrow 0} (I + \bar{b}A_\lambda^{-1})(u_0 + \bar{b}g_1^\infty) = \lim_{\lambda \downarrow 0} u_\lambda^\infty$$

Proof of Theorem 3.3. We first establish the results with A replaced by $A_{(\lambda)} = \omega I + B_\lambda$, $\lambda > 0$, where B_λ is the Yosida approximation of B, defined by $B = A - \omega I$. Note that B is m-accretive in X. Let u_λ be the strong solution of the approximating equation to (V_g) written in the form:

$$u_\lambda + \omega b * u_\lambda + \omega b * \omega^{-1} B_\lambda u_\lambda = u_0 + \omega b * \omega^{-1} g \tag{3.15}$$

Since the kernel $b \notin L^1(0,\infty)$, we transform this equation into a form which has the property that its new kernel will be in $L^1(0,\infty)$ and completely positive. Indeed, if we take the convolution of (3.15) by $r(\omega b)$ subtract the result from (3.15), and use the definition of $r(\omega b)$ we get the approximating equation equivalent to (3.15):

$$u_\lambda + r(\omega b) * \omega^{-1} B_\lambda u_\lambda = u_0 + r(\omega b) * (\omega^{-1} g - u_0) \tag{3.16}$$

From Proposition 2.1, $r(\omega b)$ is completely positive and $r(\omega b) \in L^1(0,\infty)$, with $\int_0^\infty r(\omega b)(\delta)d\delta = 1$.

To prove Theorem 3.3, part 1, we wish to apply Theorem 3.2, part 1, to (3.16). If g satisfies the assumptions of Theorem 3.3, part 1, so does $\omega^{-1}g - u_0$. Thus all assumptions of Theorem 3.2, part 1 are satisfied with b replaced by $r(\omega b)$, A replaced by $\omega^{-1}B_\lambda$, g replaced by $\omega^{-1}g - u_0$, and u by u_λ. We obtain (by (3.4)):

$$\begin{aligned} \| u_\lambda(t) - u_\lambda^\infty \| &\leq \int_t^\infty r(\omega b)(\tau)d\tau \, \| u_0 - u_\lambda^\infty \| \\ &+ (\omega^{-1}(r(\omega b) * \| g - g^\infty \|))(t) \qquad (0 \leq t < \infty) \end{aligned} \tag{3.17}$$

where u_λ^∞ is the unique solution of the limiting equation

$$u_\lambda^\infty + \omega^{-1}B_\lambda u_\lambda^\infty = \omega^{-1}g_\infty \tag{3.18}$$

which exists because B_λ is m-accretive and $\omega > 0$. Note that (3.17) is the estimate (3.16) with u replaced by u_λ.

If g satisfies the assumptions of Theorem 3.3, part 2, and $b \in AC[0,\infty]$, it follows from Proposition 2.1 that $r(\omega b) \in L^1(0,\infty) \cap C[0,\infty]$. Thus with $g = g_1 + g_2$ we can apply Theorem 3.2, part 2, to (3.16) and we obtain (from (3.5)) the estimate:

$$\begin{aligned}\| u_\lambda(t) - u_\lambda^\infty \| &\le \int_t^\infty r(\omega b)(\tau)d\tau \, \| u_0 - u_\lambda^\infty \| \\ &+ (\omega^{-1}r(\omega b)*\| g_1 - g_1^\infty \|)(t) + (\omega^{-1}r(\omega b)*\| g_2 \|)(t) \\ &(0 \le t < \infty)\end{aligned} \tag{3.19}$$

where u_λ^∞ is the unique solution of the limiting equation

$$u_\lambda^\infty + \omega^{-1}B_\lambda u_\lambda = \omega^{-1}g_1^\infty \tag{3.20}$$

Note that (3.19) is the estimate (3.7) with u replaced by u_λ.

Since B is m-accretive $\lim_{\lambda\downarrow 0} u_\lambda^\infty = u^\infty$, where in the case of (3.18) u^∞ satisfies the limiting equation

$$u^\infty + \omega^{-1}Bu^\infty \ni \omega^{-1}g^\infty \tag{3.21}$$

or equivalently u^∞ satisfies the limiting equation

$$Au^\infty \ni g^\infty \tag{3.22}$$

Similarly, in the case of (3.20) we find that u^∞ satisfies the limiting equation

$$Au^\infty \ni g_1^\infty \tag{3.23}$$

It remains to prove that $\lim_{\lambda\downarrow 0} u_\lambda = u$ in $L^1(0,T;X)$ for every $T > 0$, where u is the generalized solution of (V_g). Having done so it is immediate that the estimates (3.17), (3.19) hold with u_λ replaced by u thus (3.6) and (3.7) hold. We know that $\lim_{\lambda\downarrow 0} \int_0^T \| \tilde{u}_\lambda(t) - u(t) \| dt = 0$, where $\tilde{u}_\lambda \in L^1_{loc}(0,\infty;X)$ satisfies

$$\tilde{u}_\lambda + b*A_\lambda \tilde{u}_\lambda = u_0 + b*g \tag{3.24}$$

Introduce the notation $\delta = \lambda(1 + \lambda\omega)^{-1}$. Since $A = \omega I + B$, one easily checks that

$$A_\lambda = \omega(1 + \omega\lambda)^{-1}I + B_\delta$$

Thus the solution $\tilde{u}_\lambda$ of (3.24) satisfies the equation

$$\tilde{u}_\lambda + b*B_\delta \tilde{u}_\lambda = u_0 + b*(g - \omega(1 + \omega\lambda)^{-1}\tilde{u}_\lambda) \qquad (3.25)$$

or by (3.15):

$$u_\delta + b*B_\delta u_\delta = u_0 + b*(g - \omega u_\delta) \qquad (3.26)$$

To compare solutions of (3.25) and (3.26) we apply the inequality (3.3) of Theorem 3.1, and we obtain (note that $u_{01} = u_{02} = u_0$)

$$\| \tilde{u}_\lambda - u_\delta \| \leq \omega b*(1 + \lambda\omega)^{-1} \| \tilde{u}_\lambda - u_\delta \| + \lambda\omega^2(1 + \lambda\omega)^{-1}b*\| \tilde{u}_\lambda \|$$

and hence also the estimate

$$\| \tilde{u}_\lambda - u_\delta \| \leq \omega b*\| \tilde{u}_\lambda - u_\delta \| + \delta\omega^2 b*\| \tilde{u}_\lambda \| \qquad (3.27)$$

It follows from (3.27) that for every $T > 0$

$$\int_0^T \| \tilde{u}_\lambda - u_\delta \|(t)dt \leq \delta\omega^2 \| b \|_{L^1(0,T)} \cdot (1 + \| r(-\omega b) \|_{L^1(0,T)}) \int_0^T \| \tilde{u}_\lambda(t) \|dt \qquad (3.28)$$

Since $\tilde{u}_\lambda$ converges to u in $L^1(0,T;X)$, we finally obtain $\lim_{\lambda\downarrow 0} \int_0^T \| u_\delta - u \|(t)dt = 0$, which proves that the u_λ converges in $L^1[0,T;X]$ to the generalized solution u of (V_g) for every $T > 0$. This completes the proof of Theorem 3.3.

EXAMPLE 3.4. *A Conservation Law with Memory.* As an illustration of Theorems 3.1 and 3.2 we consider the existence and qualitative properties of positive solutions of the problem

$$u(t,x) + \int_0^t b(t - s)\phi(u(s,x))_x ds = u_0(x) \qquad (t \geq 0,\ x \in R) \qquad (c)$$

We assume that $\phi \in C^1(R)$ is a given function. If $b \equiv 1$ problem (c) is equivalent to the nonlinear conservation law in one space dimension:

$$u_t + \phi(u)_x = 0, \qquad u(0,x) = u_0(x) \qquad (x \in R)$$

Although no particular physical significance is claimed for (c), it evidently contains the usual conservation law as a special case. The latter has been studied extensively from special points of view. Crandall [24] has shown that if $\phi: R \to R$ is a given smooth strictly increasing function (actually ϕ continuous is sufficient) such that $\phi(0) = 0$, then the operator A defined by $Au = \phi(u)_x$ on the Banach space $X = L^1(R)$, with

$D(A) = \{u \in L^1(R): \phi(u_x) \in L^1(R)\}$ (see [24, Definition 1.1 and Theorem 1.1]), is m-accretive on X, and $\overline{D(A)} = X$. Moreover, one has: $J_\lambda(0) = 0$ and $J_\lambda u \leq J_\lambda v$ $(\lambda > 0)$, whenever $u \leq v$, u, $v \in L^1(R)$.

In (c) assume that $b \in L^1_{loc}(0,\infty)$, $b \not\equiv 0$ completely positive on $[0,\infty)$, and $u_0 \in \overline{D(A)}$; to be specific take b nonnegative, nonincreasing and log b convex on $(0,\infty)$, $b \not\equiv 0$. Then by Gripenberg's result [34], see Theorem 3.7, Section 2, and Remark (i) following Theorem 2.2, and by Theorem 3.1, problem (c) has a unique generalized solution u; u is nonnegative whenever u_0 is nonnegative, and $u_1 \geq u_2$ whenever $u_{01} \geq u_{02}$. If, in addition, $b \in L^1(0,\infty)$, then this generalized solution u converges strongly in $L^1(R)$ as $t \to \infty$ to the element $u_\infty \in D(A)$ which is the unique solution of the limit equation

$$u_\infty(x) + \left(\int_0^\infty b(t)dt\right) \phi(u_\infty(x))_x = u_0(x) \qquad (x \in R)$$

u_∞ exists and is uniquely defined since $\int_0^\infty b(t)dt > 0$ and A is m-accretive.

4.4. Application to Nonlinear Heat Flow in a Material with Memory

In this section we apply the theory developed in 4.2 and 4.3 of this Section to discuss the global existence, uniqueness, positivity, and decay of positive solutions of the nonlinear heat flow problem formulated in Section 1.2. This problem was discussed from a different point of view in Section 3.4. In Section 3.4, boundedness and decay results were obtained without consideration of the positivity of the data. Here the principal concern is the existence and decay of positive solutions when the data are positive.

Referring to Section 1.2, we again study the heat flow initial-boundary value problem (see (2.3) Section 1, also (4.1) Section 3):

$$\begin{cases} \frac{\partial}{\partial t}[b_0 u(t,x) + (\beta * u)(t,x)] = c_0 \sigma(u_x(t,x))_x - (\gamma * \sigma(u_x)_x)(t,x) \\ \qquad + h(t,x) \qquad (0 < t < \infty,\ 0 < x < 1) \\ u(t,0) = u(t,1) \equiv 0 \qquad (t \geq 0) \\ u(0,x) = u_0(x) \qquad (0 < x < 1) \end{cases} \tag{4.1}$$

As before we assume that the constants b_0, c_0 are positive, that $\beta, \gamma \in L^1(0,\infty)$, and that

$$b_0 + \int_0^\infty \beta(t)dt > 0 \tag{β}$$

and

$$c_0 - \int_0^\infty \gamma(t)dt > 0 \qquad (\gamma)$$

We also assume that the function $\sigma : R \to R$ satisfies assumptions (σ) of Lemma 2.3, Section 1, and that the external heat supply satisfies (at least) the assumption $h \in L^1_{loc}(0,\infty;L^2(0,1))$. In terms of the functions

$$C(t) = c_0 - \int_0^t \gamma(\tau)d\tau \qquad (0 \le t < \infty) \qquad (C)$$

$$G(t,x) = b_0\, u_0(x) + \int_0^t h(\tau,x)d\tau \qquad (0 \le t < \infty,\ 0 < x < 1) \qquad (G)$$

the heat flow problem (4.1) is equivalent to (see Section 1.2) the abstract Volterra equation

$$u + \beta{*}u + C{*}Au = G \qquad (0 \le t < \infty,\ 0 < x < 1) \qquad (V_1)$$

here we have taken the constant b_0 as 1 without loss of generality. The nonlinear operator $A = \partial\phi$, defined in Lemma 2.3, Section 1, satisfies the properties established in Lemma 2.3, Section 1, provided σ satisfies assumptions (σ), and if $u_0 \in H^1_0(0,1)$, then $G(0,x) = u_0(x) \in H^1_0(0,1) = D(\phi)$. We note also that in the case of more than one space dimension, the relevant heat flow problem formulated in Section 1.2 is equivalent to the Volterra equation (V_1) with the nonlinear operator A defined in Remark 2.4. Section 1; the theory developed below applies equally well to this case.

The Volterra equation (V_1) may be written in the standard form (V_g) of Section 4.3 above by defining the resolvent kernel $r(\beta)$ of β to be the unique solution of the linear equation:

$$r(\beta) + \beta{*}r(\beta) = \beta \qquad (0 \le t < \infty) \qquad (r(\beta))$$

clearly, if $\beta \in L^1(0,\infty)$, then $r(\beta) \in L^1_{loc}(0,\infty)$ (at least). Next, define $b : [0,\infty) \to R$ by

$$b = C - r(\beta){*}C \qquad (b)$$

where C is the function defined in (C). Then the variation of constants formula shows that (V_1) is equivalent to the Volterra equation

$$u + b{*}Au = G - r(\beta){*}G \qquad (0 \le t < \infty)$$

taking $b_0 = 1$ in (G) one sees that (V_1), as well as the heat flow problem (4.1), is equivalent to the Volterra equation

$$u + b{*}Au = u_0 + 1{*}(h - r(\beta){*}h - u_0 r(\beta)) \qquad (0 \le t < \infty) \qquad (4.2)$$

The result of applying the theory of Section 4.3 on positivity and decay of solutions of (V_g) to the heat flow problem (4.1) is:

THEOREM 4.1. Let β be bounded, nonnegative, nonincreasing and convex on $[0,\infty)$. Let γ be nonnegative, nonincreasing, log convex and bounded on $[0,\infty)$. Let assumption (γ) hold, and let

$$\beta'(t) + \frac{\gamma(0)}{c_0}\beta(t) \le 0 \qquad \text{a.e. for} \qquad t \in [0,\infty) \tag{4.3}$$

Let the assumption (σ) be satisfied, and let $A = \partial\phi$ where ϕ is defined in Lemma 2.3, Section 1.

(1) If $u_0 \in L^2(0,1)$ and if the forcing function $h \in L^2_{loc}([0,\infty) \times [0,1))$, then the nonlinear Volterra equation (4.2) (equivalent to the heat flow problem (4.1)) possesses a unique strong solution u on $[0,\infty)$, such that $tu' \in L^2_{loc}(0,\infty;L^2(0,1))$; if $u_0 \in H^1_0(0,1)$, then $u' \in L^2_{loc}(0,\infty;L^2(0,1))$.

(2) If the data u_0 and h satisfy $u_{0,1}(x) \le u_{0,2}(x)$ a.e. on $[0,1]$ and if $h^1(t,x) \le h^2(t,x)$ a.e. on $[0,\infty) \times [0,1]$; then the corresponding strong solutions u_i $(i = 1,2)$ satisfy $u_1(t,x) \le u_2(t,x)$ a.e. $[0,\infty) \times [0,1]$; in particular, if $u_0(x) \ge 0$ and $h(t,x) \ge 0$ a.e. on $[0,1]$ and $[0,\infty) \times [0,1)$ respectively, then $u(t,x) \ge 0$ a.e. on $[0,\infty) \times [0,1]$.

(3) If, in addition, $\beta \in L^1(0,\infty)$, and if $h = h_1 + h_2$, where $h_1 \in L^\infty(0,\infty;L^2(0,1))$ and there exists $h_1^\infty \in L^2(0,1)$ such that $\lim_{t\to\infty} \| h_1(t) - h_1^\infty \|_{L^2(0,1)} = 0$, and where $h_2 \in L^p(0,\infty;L^2(0,1))$ for some $p \ge 1$, then the strong solution u of [4.2] converges strongly in $L^2(0,1)$ as $t \to \infty$ to the element $u^\infty \in L^2(0,1)$; u^∞ is the unique solution of the limit equation $Au^\infty = g_1^\infty$, where

$$g_1^\infty = \frac{h_1^\infty}{c_0}\left(1 + \frac{\bar{\gamma}}{C(\infty)}\right),$$

$$\bar{\gamma} = \int_0^\infty \gamma(t)dt, \qquad C(\infty) = c_0 - \int_0^\infty \gamma(t)dt > 0$$

In particular, if $h_1^\infty = 0$, then $u^\infty = 0$.

Before giving the proof of Theorem 4.1 we state a lemma which establishes some properties of the kernel b defined by (b).

LEMMA 4.2. Let β, γ, C satisfy the assumptions of Theorem 4.1. Then b defined by (b) is completely positive on $[0,\infty)$, b satisfies the assumption

(H) (Remark (i) following Theorem 2.2), and α,k associated with b in Theorem 2.2 satisfy $\alpha = c_0^{-1} > 0$ and $k \in L^1(0,\infty)$ with

$$\int_0^\infty k(\tau)d\tau = \frac{1}{c_0}\left[\frac{1}{C(\infty)}\,\overline{\gamma}(1 + \overline{\beta}) + \overline{\beta}\right] \tag{4.4}$$

where $\overline{\beta} = \int_0^\infty \beta(t)dt$, $\overline{\gamma} = \int_0^\infty \gamma(t)dt$. Moreover, $b \notin L^1(0,\infty)$ and $b' \in L^1(0,\infty)$.

Proof. Since the functions β and $C \in AC_{loc}[0,\infty)$ it follows that the functions $r(\beta)$ and $b \in AC_{loc}[0,\infty)$ (see definitions $(r(\beta))$ and (b) respectively). Note that $b(0) = c_0 > 0$. Define $\alpha = b(0)^{-1}$ and let k be the solution of the linear Volterra equation

$$b(0)y + b'*y = \frac{b'}{b(0)} \qquad (0 \le t < \infty) \tag{k}$$

Since $b' \in L^1_{loc}[0,\infty)$, $k \in L^1_{loc}(0,\infty)$, and since

$$\frac{d}{dt}(b*k)(t) = b(0)k(t) + (b'*k)(t) = -\frac{b'(t)}{b(0)}$$

one has by integration that k satisfies the linear Volterra equation

$$\alpha b(t) + (k*b)(t) = 1 \qquad (0 \le t < \infty) \tag{4.5}$$

In order to show that b is completely positive it suffices, by Theorem 2.2, to show that k is nonnegative, nonincreasing and bounded on $[0,\infty)$. We first observe that the assumptions made on γ imply that C, -C' are convex and log (-C') is convex on $(0,\infty)$. This in turn implies that log C is convex on $(0,\infty)$, see G. Gripenberg [33]. Since C is nonnegative, nonincreasing and belongs to $L^1_{loc}(0,\infty)$, C is completely positive on $[0,\infty)$. Moreover C also satisfies assumption (H) (see Remark (i) following Theorem 2.2). It follows from Theorem 2.2 that there exists $\alpha_c > 0$, and $k_c \in L^1_{loc}(0,\infty)$, nonnegative, nonincreasing and bounded satisfying

$$\alpha_c C(t) + (k_c*C)(t) = 1 \qquad (0 \le t < \infty) \tag{4.6}$$

Note that $\alpha_c = c_0^{-1} = b^{-1}(0) = \alpha$. From the definitions of b in (b), of $r(\beta)$, and from (1.2) and (1.3) (see Section 4.1) it follows that

$$C(t) = b(t) + (\beta*b)(t) \qquad (0 \le t < \infty) \tag{4.7}$$

Substituting (4.7) into (4.6) yields

$$\alpha b + (k_c + \alpha\beta + k_c*\beta)*b = 1$$

and thus (4.5) implies that

$$k(t) = k_c(t) + \alpha\beta(t) + (k_c*\beta)(t) \qquad (0 \le t < \infty) \tag{4.8}$$

Since $k_c \in BV(0,\infty)$ we have

$$\frac{d}{dt}[\alpha\beta + k_c*\beta](t) = \alpha\beta'(t) + k_c(0)\beta(t) + \int_0^t \beta(t - \tau)dk_c(\tau) \qquad \text{a.e.}$$

on $[0,\infty)$. Hypothesis (4.3) and the identity (4.6) imply that $\alpha\beta'(t) + k_c(0)\beta(t) = 1/c_0\ [\beta'(t) + \gamma(0)/c_0\ \beta(t)] \leq 0$. Moreover, since k_c is nonincreasing and β is nonnegative

$$\int_0^t \beta(t - \tau)dk_c(\tau) \leq 0 \qquad (0 \leq t < \infty)$$

Thus k is nonnegative and nonincreasing on $[0,\infty)$. Therefore, one also has $k \in BV[0,\infty)$ if $k \in L^\infty[0,1]$. But k_c and β are bounded and $\beta \in L^1(0,1)$ imply $k \in L^\infty[0,1]$ (note that here the assumption $\beta \in L^1(0,\infty)$ is not needed). Using Theorem 2.2 again it follows that b is completely positive and satisfies assumption (H).

We next establish that $b \notin L^1(0,\infty)$. Since $C = -\gamma \in L^1(0,\infty)$, and $\lim_{t\to\infty} C(t) = C(\infty) = c_0 - \int_0^\infty \gamma(s)ds > 0$, it follows that $C \notin L^1(0,\infty)$. If $b \in L^1(0,\infty)$, it would follow from (4.7) and the assumption $\beta \in L^1(0,\infty)$ that $C \in L^1(0,\infty)$, a contradiction. Thus $b \notin L^1(0,\infty)$.

We next prove that $b' \in L^1(0,\infty]$. Indeed, from (b) it follows that

$$b'(t) = C'(t) - C(0)r(\beta)(t) - (r(\beta)*C')(t)$$

But $C' = -\gamma \in L^1(0,\infty)$; moreover, $r(\beta) \in L^1(0,\infty)$, since β is nonnegative, nonincreasing, convex and $\beta \in L^1(0,\infty)$ (use the Paley-Wiener theorem and the fact that β is positive definite).

Finally, we show that $k \in L^1(0,\infty)$. From (4.8) and the fact that $\beta \in L^1(0,\infty)$, it is sufficient to prove $k_c \in L^1(0,\infty)$. From (4.6) and the fact that C is positive, nonincreasing, $C(\infty) > 0$ and k is nonnegative, we have

$$C(\infty)\int_0^t k_c(\tau)d\tau \leq C*k_c \leq 1 \qquad (0 \leq t < \infty)$$

which proves that $k_c \in L^1(0,\infty)$. Formula (4.4) follows easily from (4.8) and (4.6). This completes the proof of Lemma 4.2.

REMARK. In Lemma 4.2, if $\beta(t) = \sum_{k=1}^n b_k e^{-\beta_k t}$ with $b_k \geq 0$ and $0 < \beta_1 < \beta_2 \ldots < \beta_n$, then condition (4.3) is satisfied if $\beta_1 \geq \gamma(0)/c_0$ holds. Indeed, since $\log \beta$ is convex and nonincreasing, it suffices to require

$$\lim_{t\to\infty} \frac{\beta'(t)}{\beta(t)} \leq -\frac{\gamma(0)}{c_0}$$

Proof of Theorem 4.1. We begin with the proof of the existence and uniqueness of strong solutions of the Volterra equation (4.2). Defining $f\colon [0,\infty) \times L^2(0,1) \to L^2(0,1)$ by

$$f = u_0 + 1*(h - r(\beta)*h - u_0 r(\beta)) \tag{4.9}$$

we have

$$f' = h - r(\beta)*h - u_0 r(\beta) \qquad (0 \le t < \infty,\ 0 < x < 1) \tag{4.10}$$

It follows from Lemma 4.2 that the kernel b satisfies the assumption

$$b(0) > 0, \qquad b \in AC_{loc}[0,\infty), \qquad b' \in BV_{loc}[0,\infty) \tag{H}$$

and that $f \in W^{1,2}_{loc}(0,\infty;L^2(0,1))$ whenever $h \in L^2_{loc}([0,\infty) \times [0,1))$. Since under assumptions (σ), $A = \partial\phi$, where ϕ is defined in Lemma 2.3, Section 1, the existence and uniqueness of strong solutions u with the properties asserted in conclusion 1 of Theorem 4.1 follows from the result of Crandall and Nohel [26, Theorem 4 and Theorems 3.5 and 3.10, Section 2].

We next establish the asymptotic results asserted in Theorem 4.1. Since $r(\beta) \in L^1(0,\infty)$ (proved in the demonstration of Lemma 4.2), and since $h = h_1 + h_2$ one has from (4.10) that

$$f' = (h_1 - r(\beta)*h_1) + (h_2 - r(\beta)*h_2) - r(\beta)u_0$$
$$(0 \le t < \infty,\ 0 < x < 1)$$

where $h_1 - r(\beta)*h_1 \in L^\infty(0,\infty;L^2(0,1))$, and

$$\lim_{t\to\infty} (h_1 - r(\beta)*h_1)(t) = (1 - \textstyle\int_0^\infty r(\beta)(\tau)d\tau)h_1^\infty$$
$$= s(\beta)(\infty)h_1^\infty = (1 + \overline{\beta})^{-1}h_1^\infty$$

moreover,

$$(h_2 - r(\beta)*h_2 - r(\beta)u_0) \in L^1(0,\infty;L^2(0,1)) \cap L^p(0,\infty;L^2(0,1))$$

with $1 < p < \infty$.

By using (4.10), the fact that the kernel b defined by (b) is by Lemma 4.2 completely positive, and Theorem 2.2, we can write the Volterra equation (4.2) in the equivalent form

$$u + b*Au = u_0 + b*(\alpha f' + k*f') \qquad (0 \le t < \infty) \tag{4.11}$$

To arrive at (4.11) we use the relation $\alpha b + k * b = 1$ in the right hand side of (4.9) and recombine terms making use of (4.10). Thus (4.11) is in the basic form (V_g) of Section 4.3 with

$$g = \alpha f' + k*f' \qquad (0 \le t < \infty) \tag{4.12}$$

From Lemma 4.2 $k \in L^1(0,\infty)$, and $g = g_1 + g_2$, where (using $h = h_1 + h_2$ in (4.10))

$$g_1 = \alpha(h_1 - r(\beta)*h_1) + k*(h_1 - r(\beta)*h_1) \quad (0 \leq t < \infty) \tag{4.13}$$

$$g_2 = g_{2,1} + g_{2,2} \quad (0 \leq t < \infty) \tag{4.14}$$

with

$$g_{2,1} = -\alpha u_0 r(\beta) - \alpha r(\beta)*h_2 + k*(h_2 - r(\beta)*h_2 - u_0 r(\beta)), \qquad g_{22} = \alpha h_2$$

Clearly $g_{21} \in L^1(0,\infty;L^2(0,1))$ and $g_{22} \in L^p(0,\infty;L^2(0,1))$. From Lemma 4.2 one has that b is completely positive on $[0,\infty)$, $b \notin L^1(0,\infty)$, and $b' \in L^1(0,\infty)$. Thus all assumptions of Theorem 3.3, part 2, are satisfied. We conclude that estimate (3.7) of Theorem 3.3, Section 4.3, holds, and therefore $\lim_{t\to\infty} \| u(t) - u^\infty \| = 0$, where $u^\infty = A^{-1}g_1^\infty$ with g_1^∞ given in the statement of Theorem 4.1; note that to evaluate g_1^∞ use is made of (4.4) and of Proposition 2.1.

Finally, we establish the "comparison" result asserted in Theorem 4.1, part 2. Let $p = \{u \in L^2(0,1) : u \geq 0\}$; p is a closed convex cone in $L^2(0,1)$ and $v - u \in P$ if and only if $u \leq v$. Moreover, it is standard that if $u \leq v$ then $J_\lambda u \leq J_\lambda v$ for every $\lambda > 0$, where $J_\lambda = (I + \lambda A)^{-1}$. We shall prove the result for solutions of the Volterra equation (V_1) which is equivalent to (4.2). As usual we shall prove the result for solutions of the approximating equation $(V_{1\lambda})$ of (V_1) in which A is replaced by the Yosida approximation A_λ, $\lambda > 0$, and then obtain the result by letting $\lambda \downarrow 0$.

Let $u_{0,i} \in L^2(0,1)$, $h_i \in L^1(0,1)$, $i = 1,2$, satisfy $u_{01} \leq u_{02}$ and $h_1(t) \leq h_2(t)$ a.e. on $[0,T]$; let $u_{\lambda,i}$ be the strong solutions of the approximating equation

$$u_{\lambda,i} + \beta*u_{\lambda,i} + \lambda^{-1}C*u_{\lambda,i} = \lambda^{-1}C*J_\lambda u_{\lambda,i} + u_{0,i} + 1*h_i \tag{$V_{1\lambda}$}$$

$$(i = 1,2;\ \lambda > 0;\ 0 \leq t \leq T)$$

It follows by an elementary calculation (which uses the definitions of b, $r(\lambda^{-1}b)$, and the relation $C \equiv b + \beta*b$) that

$$u_{\lambda,i} = r(\lambda^{-1}b)*J_\lambda u_{\lambda,i} + f_{\lambda,i} \quad (i = 1,2;\ \lambda > 0) \tag{4.15}$$

where $f_{\lambda,i}$ are solutions of the linear Volterra equation

$$f_{\lambda,i} + \beta*f_{\lambda,i} + \lambda^{-1}C*f_{\lambda,i} = u_{0i} + 1*h_i \quad (i = 1,2) \tag{4.16}$$

Hence by a familiar calculation one has

$$f_{\lambda,i} = u_{0,i}s(\beta + \lambda^{-1}C) + h_i*s(\beta + \lambda^{-1}C) \quad (i = 1,2)$$

and from (4.15), (4.16) the difference $u_{\lambda,2} - u_{\lambda,1}$ satisfies the Volterra equation

$$u_{\lambda,2} - u_{\lambda,1} = r(\lambda^{-1}b)*(J_\lambda u_{\lambda,2} - J_\lambda u_{\lambda,1}) \\ + (u_{0,2} - u_{0,1})s(\beta + \lambda^{-1}C) + (h_2 - h_1)*s(\beta + \lambda^{-1}C) \tag{4.17}$$

Since $\beta + \lambda^{-1}C$ is positive, nonincreasing, it follows from Levin's result (see Section 2.2) that $s(\beta + \lambda^{-1}C)(t) \geq 0$. Thus

$$z_\lambda = (u_{0,2} - u_{0,1})\ s(\beta + \lambda^{-1}C) + (h_2 - h_1)*s(\beta + \lambda^{-1}C)$$

satisfies $z_\lambda(t) \geq 0$ a.e. on $[0,\infty)$.

Next, define $v_\lambda = u_{\lambda,2} - u_{\lambda,1} - z_\lambda$; using (4.17) we have that v_λ satisfies

$$v_\lambda = r(\lambda^{-1}b)*(J_\lambda(v_\lambda + u_{\lambda,1} + z_\lambda) - J_\lambda(u_{\lambda,1})) \tag{4.18}$$

As in [18] (see also Remark (iii) following the statement of Theorem 3.1, Section 4.3) one shows that $v_\lambda = \lim_{n\to\infty} v_{\lambda,n}$, where

$$v_{\lambda,n+1} = r(\lambda^{-1}b)*(J_\lambda(v_{\lambda,n} + u_{\lambda,1} + z_\lambda) - J_\lambda(u_{\lambda,1})) \tag{4.19}$$

where $v_{\lambda,1} \in L^1(0,T;L^2(0,1))$ is arbitrary. Choosing $v_{\lambda,1}(t) \in p$ a.e. on $[0,T]$, $T > 0$ arbitrary, one shows easily that $v_{\lambda,n}(t) \in p$ a.e. on $[0,T]$ for all positive integers n. This also uses the fact that $r(\lambda^{-1}b) \geq 0$, that J_λ is an increasing map with respect to the ordering $\leq$, and that $z_\lambda(t) \in p$ a.e. on $[0,T]$. Thus $v_\lambda(t) \in p$ a.e. on $[0,T]$, and for $\lambda > 0$:

$$u_{\lambda,2}(t) - u_{\lambda,1}(t) = z_\lambda(t) + v_\lambda(t) \geq 0 \quad \text{a.e. on } [0,T] \tag{4.20}$$

Since $T > 0$ is arbitrary, (4.20) holds on $[0,\infty)$, and the conclusion follows by letting $\lambda \downarrow 0$. This completes the proof of Theorem 4.1.

APPENDIX

Proof of Lemma 2.2. (a) Consider the resolvent equation (k) of B'. Since $B' \in L^1(0,\infty)$, the Paley-Wiener theorem [69] yields that $k \in L^1(0,\infty)$ if and only if

$$P(z) = b(0) + \hat{B}'(z) = b_\infty + z\hat{B}(z) \neq 0 \qquad (\text{Re } z \geq 0)$$

With $z = x + iy$

$$\text{Re } P(z) = b_\infty + x \text{ Re } \hat{B}(z) - y \text{ Im } \hat{B}(z) \qquad (x \geq 0)$$

$$\text{Im } P(z) = x \text{ Im } \hat{B}(z) + y \text{ Re } \hat{B}(z) \qquad (x \geq 0)$$

Since $P(z)$ is analytic in $\text{Re } z > 0$ and continuous in $\text{Re } z \geq 0$, $\text{Re } P(z)$ and $\text{Im } P(z)$ are harmonic for $x \geq 0$. Hence by the maximum principle for harmonic functions, $P(z) \neq 0$ for $x \geq 0$ if either $\text{Re } P(iy) = b_\infty - y \text{ Im } \hat{B}(iy)$, or $\text{Im } P(iy) = y \text{ Re } \hat{B}(iy)$ are different from zero for $-\infty < y < \infty$. But by the frequency domain condition (F) $\text{Re } P(iy) > 0$ for $-\infty < y < \infty$, and thus $k \in L^1(0,\infty)$.

(b) Since $B' \in L^1(0,\infty) \cap L^2(0,\infty)$ and $k \in L^1(0,\infty)$, one has $B' * k = k * B' \in L^2(0,\infty)$, and the result $k \in L^2(0,\infty)$ follows by inspection of equation (k). If also $B'' \in L^1(0,\infty)$, then $B' \in C[0,\infty)$ (so that $|B'(0)| < \infty$) and we may differentiate (k) to obtain

$$b(0)k'(t) + B'(0)k(t) + (B''*k)(t) = -\frac{B''(t)}{b(0)} \qquad (0 \leq t < \infty)$$

and clearly $k' \in L^1(0,\infty)$.

(c) If, as is the case here $k' \in L^1(0,\infty)$, the energy inequality in (c) is derived by the following simple argument (see the method of [67, Theorem 1]). Extend k' evenly for $t < 0$, and let

$$w_T(t) = \begin{cases} w(t) & \text{if} \quad t \in [0,T] \\ 0 & \text{otherwise} \end{cases}$$

Then

(d) Multiply equation (k) by $\sqrt{t}$:

$$b(0)\sqrt{t}\, k(t) + \sqrt{t}\, (B'*k)(t) = -\frac{\sqrt{t}\, B'(t)}{b(0)} \qquad (0 \leq t < \infty)$$

An elementary calculation involving $\sqrt{t}(B'*k)$ shows that $\sqrt{t}\, k$ satisfies

$$b(0)\sqrt{t}\, k(t) + \int_0^t B'(t-\tau)\sqrt{\tau}\, k(\tau)d\tau$$

$$= -\frac{\sqrt{t}\, B'(t)}{b(0)} - \int_0^t (\sqrt{t} - \sqrt{\tau})B'(t-\tau)k(\tau)d\tau \qquad (0 \leq t < \infty)$$

Since $\sqrt{t} - \sqrt{\tau} \leq \sqrt{t-\tau}$ for $0 \leq \tau \leq t$, and since also $\sqrt{t}\, B' \in L^1(0,\infty)$ by assumption and $k \in L^1(0,\infty)$ by (a), the integral on the right side of the last equation defines a function in $L^1(0,\infty)$. Then $\sqrt{t}\, k \in L^1(0,\infty)$ by the argument of part (a). The additional assumptions and elementary estimates also yield $\sqrt{t}\, k \in L^2(0,\infty)$.

Finally, differentiating the equation (k) as in part (b), multiplying the resulting equation by $\sqrt{t}$, and using elementary estimates yields $\sqrt{t}\, k' \in L^1(0,\infty)$. This completes the proof of Lemma 2.2.

ACKNOWLEDGMENTS

Research sponsored by the United States Army under Contract Nos. DAAG29-75-C-0024 and DAAG29-80-C-0041.

I wish to express sincere appreciation to the editors: Terry Herdman, Samuel Rankin, and Harlan Stech for their encouragement to prepare these notes, and to Professor Rankin for organizing a stimulating and timely conference. Finally, I wish to thank Mrs. Sally Ross for painstaking preparation of the original manuscript.

REFERENCES

1. S. Aizicovici, On a nonlinear integrodifferential equation, *J. Math. Anal. Appl.* 63 (1978), 385-395.

2. S. Aizicovici, Existence theorems for a class of integrodifferential equations, *An. Sti. Univ. "Al. I. Cuza,"* Iasi (to appear).

3. S. Aizicovici, On an abstract Volterra equation. *Volterra Equations*, Lecture Notes in Mathematics No. 737, Springer Verlag (1979), 1-8.

4. H. Attouch and A. Damlamian, A nonlinear Volterra equation in variable domain, Mathematics Research Center Technical Summary Report #2028, University of Wisconsin-Madison, September 1979.

5. V. Barbu, Integro-differential equations in Hilbert spaces, *An. Sti. Univ. "Al. I. Cuza,"* Iasi Sect. I a Mat. 19 (1973), 365-383.

6. V. Barbu, Nonlinear Volterra equations in a Hilbert space, *SIAM J. Math. Anal.* 6 (1975), 728-741.

7. V. Barbu, *Nonlinear Semigroups and Differential Equations in Banach Spaces*, Noordhoff, Groningen, 1976.

8. V. Barbu, On a nonlinear Volterra integral equation on a Hilbert space, *SIAM J. Math. Anal.* 8 (1977), 345-355.

9. V. Barbu, Existence for nonlinear Volterra equations in Hilbert spaces, *SIAM J. Math. Anal.* 9 (1978).

10. V. Barbu, Degenerate nonlinear Volterra integral equations in Hilbert space. *Volterra Equations*, Lecture Notes in Mathematics No. 737, Springer Verlag (1979), 9-21.

11. V. Barbu and M. A. Malik, Semilinear integrodifferential equations in Hilbert spaces, *J. Math. Anal. Appl.* (to appear).

12. R. Bellman and K. L. Cooke, *Differential-Difference Equations*, Academic Press, 1963.

13. Ph. Benilan, Equations d'évolution dans une espace de Banach quelconque et applications, Thèse de Doctorat d'État, Univ. de Paris (Orsay), 1972.

14. H. Brézis, *Operateurs Maximaux Monotones et Semi-Groupes de Contractions dans les Espaces de Hilbert*, North Holland, 1973.

15. H. Brézis, Monotonicity Methods in Hilbert Spaces and Some Applications to Nonlinear Partial Differential Equations, *Contributions to Nonlinear Functional Analysis*, Academic Press, New York, 1971, pp. 101-156.

16. R. W. Carr and K. B. Hannsgen, A nonhomogeneous integrodifferential equation in Hilbert space, *SIAM J. Math. Anal.* 10 (1979), 961-983.

17. Ph. Clément, On abstract Volterra equations with kernels having a positive resolvent, *Israel J. Math.* (to appear) (Mathematics Research Center Technical Summary Report #1979, University of Wisconsin-Madison, July, 1979).

18. Ph. Clément and J. A. Nohel, Abstract linear and nonlinear Volterra equations preserving positivity, *SIAM J. Math. Anal.* 10 (1979), 365-388.

19. Ph. Clément and J. A. Nohel, Asymptotic behavior of positive solutions of nonlinear Volterra equations for heat flow, Mathematics Research Center Technical Summary Report #2069, University of Wisconsin-Madison, 1980 (to appear).

20. Ph. Clément, R. C. MacCamy, and J. A. Nohel, Asymptotic properties of solutions of nonlinear abstract Volterra equations, Mathematics Research Center Technical Summary Report #2070, University of Wisconsin-Madison, 1980 (to appear).

21. B. D. Coleman, Thermodynamics of materials with memory, *Arch. Rational Mech. Anal.* 17 (1964), 1-46.

22. B. D. Coleman and M. E. Gurtin, Equipresence and constitutive equations for rigid heat conductors, *Z. Angew. Math. Phys.* 18 (1967), 199-207.

23. B. D. Coleman and V. J. Mizel, Thermodynamics and departures from Fourier's law of heat conduction, *Arch. Rational Mech. Anal.* 13 (1963), 245-261.

24. M. G. Crandall, The semigroup approach to first order quasilinear equations in several space variables, *Israel J. Math.* 12 (1972), 108-132.

25. M. G. Crandall, An Introduction to Evolution Governed by Accretive Operators, in *Dynamical Systems*, Vol. 1, Academic Press, 1976, pp. 131-165.

26. M. G. Crandall and J. A. Nohel, An abstract functional differential equation and a related nonlinear Volterra equation, *Israel J. Math.* 29 (1978), 313-328.

27. M. G. Crandall and A. Pazy, Semigroups of nonlinear contractions and dissipative sets, *J. Functional Analysis* 3 (1969), 376-418.

28. M. G. Crandall, S. O. Londen, and J. A. Nohel, An abstract nonlinear Volterra integrodifferential equation, *J. Math. Anal. Appl.* 64 (1978), 701-735.

29. C. M. Dafermos and J. A. Nohel, Energy methods for nonlinear hyperbolic Volterra integrodifferential equations, *Comm. in P.D.E.* 4 (1979), 219-278.

30. G. Gripenberg, An existence result for a nonlinear Volterra integral equation in a Hilbert space, *SIAM J. Math. Anal.* 9 (1978), 793-805.

31. G. Gripenberg, On a nonlinear Volterra integral equation in a Banach space, *J. Math. Anal. Appl.* 66 (1978), 207-219.

32. G. Gripenberg, On some integral and integrodifferential equations in a Hilbert space, *Annali di Mat. Pura ed Appl.*(IV) 118 (1978), 181-198.

33. G. Gripenberg, On positive, nonincreasing resolvents of Volterra equations, *J. Differential Equations* 30 (1978), 380-390.

34. G. Gripenberg, An abstract nonlinear Volterra equation, *Israel J. Math.* 34 (1979), 198-212.

35. G. Gripenberg, On a frequency domain condition used in the theory of Volterra integral equations, *SIAM J. Math. Anal.* 10 (1979), 839-843.

36. G. Gripenberg, On an abstract integral equation, *SIAM J. Math. Anal.* 10 (1979), 1017-1021.

37. G. Gripenberg, On nonlinear Volterra equations with nonconvolution kernels in abstract spaces, Helsinki U. of Tech. Report HTKK-MAT-A105 (1977).

38. G. Gripenberg, On Volterra equations with nonconvolution kernels, Helsinki U. of Tech. Report HTKK-MAT-A118 (1978).

39. M. E. Gurtin and A. C. Pipkin, A general theory of heat conduction with finite wave speeds, *Arch. Rat. Mech. Anal.* 31 (1968), 113-126.

40. A. Halanay, On the asymptotic behavior of solutions of an integrodifferential equation, *J. Math. Anal. Appl.* 10 (1965), 319-324.

41. K. B. Hannsgen, Indirect Abelian theorems and a linear Volterra equation, *Amer. Math. Soc.* 142 (1969), 539-555.

42. K. B. Hannsgen, A Volterra equation with completely montonic convolution kernel, *J. Math. Anal. Appl.* 31 (1970), 459-471.

43. K. B. Hannsgen, A Volterra equation with parameter, *SIAM J. Math. Anal.* 4 (1973), 22-30.

44. K. B. Hannsgen, A Volterra equation in Hilbert space, *Ibid.* 5 (1974), 412-416.

45. K. B. Hannsgen, A linear Volterra equation in Hilbert space, *Ibid.* 5 (1974), 927-940.

46. K. B. Hannsgen, The resolvent kernel of an integrodifferential equation in Hilbert space, *Ibid.* 7 (1976), 481-490.

47. K. B. Hannsgen, Uniform L^1 behavior for an integrodifferential equation with parameter, *Ibid.* 8 (1977), 626-639.

48. T. Kiffe and M. Stecher, L^2 solutions of Volterra integral equations, *SIAM J. Math. Anal.* 10 (1979), 274-280.

49. T. Kiffe, A perturbation of an abstract Volterra equation, *SIAM J. Math. Anal.* (to appear).

50. P. D. Lax, Development of singularities of solutions of nonlinear hyperbolic partial differential equations, *J. Math. Phys.* 5 (1964), 611-613.

51. J. J. Levin, On a nonlinear Volterra equation, *J. Math. Anal. Appl.* 39 (1972), 458-476.

52. J. J. Levin, Resolvents and bounds for linear and nonlinear Volterra equations, *Trans. Amer. Math. Soc.* 228 (1977), 207-222.

53. S. -O. Londen, On a nonlinear Volterra equation, *J. Differential Equations* 14 (1973), 106-120.

54. S. -O. Londen, On an integral equation in a Hilbert space, *SIAM J. Math. Anal.* 8 (1977), 950-970.

55. S. -O. Londen, An existence result for a Volterra equation in Banach space, *Trans. Amer. Math. Soc.* 235 (1978), 285-305.

56. S. -O. Londen, An integrodifferential Volterra equation with a maximal monotone mapping, *J. Differential Equations* 27 (1978), 405-420.

57. R. C. MacCamy, Approximations for a class of functional differential equations, *SIAM J. Appl. Math.* 23 (1972), 70-83.

58. R. C. MacCamy, Stability theorems for a class of functional differential equations, *SIAM J. Appl. Math.* 30 (1976), 557-576.

59. R. C. MacCamy, An integro-differential equation with applications in heat flow, *Quarterly Appl. Math.* 35 (1977), 1-19.

60. R. C. MacCamy, A model for one-dimensional, nonlinear viscoelasticity, *Ibid.* 35 (1977), 21-33.

61. R. C. MacCamy and P. Weiss, Numerical approximations for Volterra integral equations. *Volterra Equations*, Lecture Notes in Mathematics No. 737, Springer Verlag (1979), 173-191.

62. R. K. Miller, On Volterra integral equations with nonnegative integrable resolvents, *J. Math. Anal. Appl* 22 (1968), 319-340.

63. R. K. Miller, *Nonlinear Volterra Integral Equations*, W. A. Benjamin, Menlo Park, CA, 1971.

64. J. Naumann, On a class of first order evolution inequalities arising in heat conduction with memory, *SIAM J. Math. Anal.* 10 (1979), 1144-1160.

65. T. Nishida, Global smooth solutions of the second-order quasilinear wave equations with the first-order dissipation (unpublished).

66. J. A. Nohel, A forced quasilinear wave equation with dissipation. *EQUADIFF* 4, Lecture Notes in Mathematics No. 703, Springer Verlag (1977), 318-327.

67. J. A. Nohel and D. F. Shea, Frequency domain methods for Volterra equations, *Advances in Math.* 22 (1976), 278-304.

68. J. W. Nunziato, On heat conduction in materials with memory, *Quarterly Appl. Math.* 29 (1971), 187-304.

69. R. E. A. C. Paley and N. Wiener, *Fourier Transforms in the Complex Domain*, Amer. Math. Soc. Colloquium Publications, 1934.

70. A. Pazy, Strong convergence of semigroups of nonlinear contractions in Hilbert space, *J. d'Analyse Mathém.* 34 (1978), 1-35.

71. C. Rennolet, Existence and boundedness of abstract nonlinear integro-differential equations of nonconvolution type, *J. Math. Anal. Appl.* 70 (1979), 42-60.

72. C. Rennolet, Existence and boundedness results for abstract nonlinear Volterra equations of nonconvolution type, *J. Integral Equations* (to appear).

73. D. F. Shea and S. Wainger, Variants of the Wiener-Lévy theorem, with applications to stability problems for some Volterra integral equations, *Amer. J. Math.* 97 (1975), 312-343.

74. O. J. Staffans, Nonlinear Volterra equations with positive definite kernels, *Proc. Amer. Math. Soc.* 5 (1975), 103-108.

75. O. J. Staffans, Positive definite measures with applications to a Volterra equation, *Trans. Amer. Math. Soc.* 218 (1976), 219-237.

76. O. J. Staffans, Tauberian theorems for a positive definite form with applications to a Volterra equation, *Trans. Amer. Math. Soc.* 218 (1976), 239-259.

77. O. J. Staffans, An inequality for positive definite kernels, *Proc. Amer. Math. Soc.* 58 (1976), 205-210.

78. O. J. Staffans, Boundedness and asymptotic behavior of solutions of a Volterra equation, *Mich. Math. J.* 24 (1977), 77-95.

79. O. J. Staffans, Some energy estimates for a nondifferentiated Volterra equation, *J. Differential Equations* 32 (1979), 285-293.

80. O. J. Staffans, On a nonlinear hyperbolic Volterra equation, *SIAM J. Math. Anal.* 11 (1980), 793-812.

81. O. J. Staffans, A Volterra equation with square integrable solution, *Proc. Amer. Math. Soc.* 78 (1980), 213-217.

82. M. Tangredi, Stability properties of kernels of Volterra equations and their resolvents with applications, Ph.D. Thesis (in preparation), University of Wisconsin-Madison, 1980.

83. I. I. Vrabie, Compactness methods for an abstract nonlinear Volterra integrodifferential equation, *Nonlinear Anal., Theory, Meth. and Appl.* (to appear).

Part II: INVITED LECTURES

NONLOCAL BOUNDARY CONDITIONS

R. C. MacCamy

Department of Mathematics
Carnegie-Mellon University
Pittsburgh, Pennsylvania

1. INTRODUCTION

The term nonlocal boundary condition is best understood by contrasting it to the usual situation of local conditions. The typical boundary-value problem for a second order elliptic equation specifies the solution u or its normal derivative u_n or some combination $u_n + \beta u$ at each point of the boundary. In these situations only values at the given point are required. Here we consider problems in which u or u_n are functionals of the solution.

The author was led to the unusual conditions discussed here during a search for variational procedures for the solution of a class of problems in electromagnetic theory. In these problems there is an ambient region (usually air) and obstacles which may be either metal or non-conducting and in fact are often inhomogeneous. Different equations hold in the ambient region and in the obstacles and there are continuity requirements across the interfaces.

The class of problems just described is complicated by the fact that the ambient region is assumed infinite in extent. On the other hand variational principles, if they are to be implemented numerically, must be restricted to bounded regions. We have succeeded in finding two such principles. In each case the price paid for the reduction to finite regions is the appearance of nonlocal boundary conditions.

We illustrate the above ideas here with a model problem. Most of the technical difficulties associated with the electromagnetic problems have been eliminated but the essential notions are illustrated. We let Γ denote a simple closed curve in the plane with exterior A and interior B. Let $u^o(x)$, $x = (x_1,x_2)$, satisfy $\Delta u^o - u^o = 0$ in R^2. Then we seek u(x) such that

$$\Delta u - u = 0 \quad \text{in} \quad A, \qquad \Delta u(x) - q(x)u(x) = 0 \quad \text{in} \quad B$$

$$u^+ = u^-, \qquad u_n^+ = u_n^- \quad \text{on} \quad \Gamma \tag{P}$$

$$u - u^o \to 0 \quad \text{as} \quad |x| \to \infty$$

Here n is the normal to Γ and the plus and minus signs denote limits from A and B respectively. q(x) is assumed positive in $\overline{B}$.

Our goal, again, is to obtain variational formulations for (P), these principles involving only finite regions. There are also two other requirements. We demand that boundary conditions in the principles should be natural and that the principle admit a numerical approximation procedure yielding algebraic equations with sparse matrices.

The first method we present is a specialization of work with S. P. Marin, details of which appear in [7]. It is a variation of the common engineering practice of introducing artificial boundaries and leads to a problem in a region A_∞ which is finite but larger than B. The second method evolved from work [8] with J. H. McWhirter of Westinghouse Corporation and was suggested originally by an idea in [2]. The second method is more complicated than the first but results in a variational principle over the region B itself.

The principles are described in Section 2. We make a few remarks there on numerical implementation but we do not discuss this aspect in detail since it requires considerable technical machinery. In Section 3 we outline very briefly the essential ideas in a verification that our variational problems are well posed. Although we do not discuss this those same ideas can be used to establish the convergence of finite element approximations to both principles.

2. THE VARIATIONAL PROCEDURES

METHOD I. We begin with an auxiliary problem. We enclose B with a simple closed curve Γ_∞, for instance a circle. We let Ω_∞ denote the exterior of Γ_∞ and A_∞ the annular region between Γ and Γ_∞. Then we consider the problem:

$$\begin{aligned} &\Delta v - v = 0 \quad \text{in} \quad \Omega_\infty, \quad v = \phi \quad \text{on} \quad \Gamma_\infty \\ &\quad v \to 0 \quad \text{as} \quad |x| \to \infty \end{aligned} \tag{P_∞}$$

The solution, assuming it exists, is linear in ϕ and we can therefore define a linear operator T by the formula,

$$T[\phi] = v_n^+ \quad \text{on} \quad \Gamma_\infty \tag{2.1}$$

We will indicate a little later how to solve (P_∞). First, however, we indicate how to use the operator T. Suppose u is a solution of (P). Then $u - u^o$ is a solution of (P_∞) with $\phi = (u - u^o)|_{\Gamma_\infty}$. Moreover $u - u^o$ and $\partial/\partial n\,(u - u^o)$ are continuous across Γ_∞. Hence if $w = u - u^o$ then w will satisfy,

$$\begin{aligned} &\Delta w - w = 0 \quad \text{in} \quad A_\infty \\ &\Delta w - qw = (1 - q)u^o \quad \text{in} \quad B \\ &w^+ = w^-, \quad w_n^+ = w_n^- \quad \text{on} \quad \Gamma \\ &w_n^- = T[w] \quad \text{on} \quad \Gamma_\infty \end{aligned} \tag{P_I}$$

We have now a problem on the finite region $\overline{B} \cup A_\infty$ but the boundary condition on Γ_∞ is nonlocal. We can give a variational formulation of (P_I). Let H denote the Hilbert space,

$$H = \{w : w|_B \in H_1(B),\ w|_{A_\infty} \in H_1(A_\infty),\ w^+ = w^- \quad \text{on} \quad \Gamma\}$$

with the norm,

$$\| w \|_H^2 = \| w \|_{1,B}^2 + \| w \|_{1,A_\infty}^2$$

Define the bilinear form $a(w,\phi)$ on $H \times H$ by,

$$a(w,\phi) = - \int_B (\nabla w \cdot \nabla \phi + qw\phi)dx - \int_{A_\infty} (\nabla w \cdot \nabla \phi + w\phi)dx + \int_{\Gamma_\infty} T[w]\phi \, ds \tag{2.2}$$

and the functional F by

$$F(\phi) = \int_B (1 - q)u^o \phi \, dx$$

Then the desired variational problem is: Find $w \in H$ such that,

$$a(w,\phi) = F(\phi) \quad \text{for all} \quad \phi \in H \tag{V_I}$$

If w is a solution of (P_I) then multiplication by ϕ and integration by parts yields (V_I). Conversely if w satisfies (V_I) and has enough smoothness then integration by parts in the other direction shows that w solves (P_I).

We return to the determination of T. It is shown in [5] that (P_∞) can be solved in the following manner. Define $V(x;f)$ for $x \in \Omega_\infty$ by

$$V(x;f) = \int_{\Gamma_\infty} f(y)G(x,y)ds_y \tag{2.3}$$

and $V_o(x_o;f)$ for $x_o \in \Gamma_\infty$ by,

$$V_o(x_o;f) = \int_{\Gamma_\infty} f(y)G(x_o,y)ds_y \tag{2.4}$$

Here $G(x,y) = K_o(|x - y|)$ where $K_o(z)$ is the singular Bessel function of second kind with imaginary argument. Thus $K_o(z) \sim (2\pi)^{-1}\log z$ as $z \to 0$ and $K_o(z) = O(e^{-z})$ as $z \to \infty$. The solution procedure from [5] is then as follows. Find f such that,

$$V_o(x_o;f) = \phi \quad \text{on} \quad \Gamma_\infty \tag{2.5}$$

Then $v(x) = V(x;f)$ solves (P_∞).

Once f is determined from (2.5) then (2.3) and the jump relations from potential theory yield,

$$\begin{aligned}(\frac{\partial v}{\partial n}(x_o))^+ &= \frac{1}{2} f(x_o) + \int_{\Gamma_\infty} f(y) \frac{\partial}{\partial n_x} G(x_o,y)ds_y \\ &= (\frac{1}{2}I + M)\ [f]\ (x_o)\end{aligned} \tag{2.6}$$

Details of the solution of (2.5) appear in [6]. It follows from results in that reference that the solution of (2.5) yields a map $V_o^{-1} : \phi \to f$ from $H_{1/2}(\Gamma_\infty)$ into $H_{-1/2}(\Gamma_\infty)$. On the other hand, for a smooth curve Γ_∞, the normal derivative in (2.6) is a smooth function and M maps $H_{1/2}(\Gamma_\infty)$ into itself. Thus we have,

$$T[\phi] = (\frac{1}{2}I + M)V_o^{-1}[\phi] \tag{2.7}$$

as a map from $H_{1/2}(\Gamma_\infty)$ into $H_{-1/2}(\Gamma_\infty)$.

We observe that problem (V_I) satisfies the first of our requirements in that the boundary condition is natural, that is one need not impose any conditions on the solution or trial functions. It also satisfies the second requirement as we explain now very briefly.

In order to implement (V_I) we choose a sequence of finite dimensional subspaces S^h of H depending on a parameter h. We let $\phi_1^h,\ldots,\phi_{N^h}^h$ be a basis for S^h. Then we consider the approximate variational problems: Find $w^h \in S^h$ such that

$$a(w^h,\phi_k^h) = F(\phi_k^h), \quad k = 1,\ldots,N^h \tag{V_I^h}$$

If we put $w^h = \sum_1^{N^h} w_j^h \phi_j^h$ then (V_I^h) yields the algebraic equations,

$$\sum_1^{N^h} w_j^h a(\phi_j^h, \phi_k^h) = F(\phi_k^h), \qquad k = 1,\ldots,N^h \tag{2.8}$$

In order to make the matrices in (2.8) sparse one chooses bases for S^h which are piecewise polynomials with small support, the finite element idea.

METHOD II. In this method we seek the solution of (P) in the form,

$$u(x) = \int_B G(x,y)\psi(y)dy + u^o(x) \tag{2.9}$$

where G is as in (2.3). This satisfies all the conditions of (P) save for the equation in B. From (2.9) we have $\Delta u - u = \psi$ in B hence we will satisfy the equation in B if,

$$\psi(x) = (q - 1)(\int_B G(x,y)\psi(y)dy + u^o(x)) \tag{2.10}$$

The integral equation (2.10) is the result of [2] in the present situation. We could give an equivalent variational problem for it but it would violate our second requirement. The problem is that the integral operator is nonlocal hence the introduction of finite elements would produce full matrices. Thus we perform a further reduction.

Let ψ be a solution of (2.10). Then we calculate as follows:

$$\Delta(\frac{\psi}{q-1}) = \Delta(\int_B G(x,y)\psi(y)dy + u^o) = \psi + (\int_B G\psi dy + u^o)$$

$$= \psi + \frac{\psi}{q-1} = q(\frac{\psi}{q-1})$$

Hence if $\psi = (q - 1)w$ then,

$$\Delta w - qw = 0 \quad \text{in} \quad B \tag{2.11}$$

Now we reverse this calculation. Let w be a solution of (2.11) and put $\psi = (q - 1)w$. Consider then the function χ defined by

$$\chi(x) = w - \int_B G(x,y)\psi(y)dy - u^o(x)$$

We have,

$$\Delta\chi = \Delta w - \psi - \int_B G(x,y)\psi(y)dy - u^o(x) = qw - (q - 1)w$$

$$- \int_B G(x,y)\psi(y)dy - u^o(x) = \chi$$

Thus if $\chi \equiv 0$ on Γ then $\chi \equiv 0$ in all of B, that is (2.10) holds.

We have thus been led to a new problem. Put,

$$V[w](x_o) = \int_B G(x,y)(q(y) - 1)w(y)dy \tag{2.12}$$

Then the new problem is: Find w such that,

$$\begin{aligned} &\Delta w - qw = 0 \quad \text{in} \quad B \\ &w(x_o) = V[w](x_o) + u^o(x) \quad \text{on} \quad \Gamma \end{aligned} \tag{P_{II}}$$

The calculation above shows that if w is a solution of (P_{II}) and $\psi = (q - 1)w$ then ψ is a solution of (2.10) and hence (2.9) gives a solution of (P).

We now give a variational formulation of (P_{II}). In order that the boundary conditions be natural we use a hybrid method following the idea in [4]. Let w be a solution of (P_{II}) and ϕ be a smooth function. Then we have on integration by parts,

$$- \int_B(\nabla w \cdot \nabla \phi + qw\phi)dx + \int_\Gamma w_n \phi \, ds = 0 \tag{2.13}$$

Also if μ is a smooth function on Γ then the boundary condition yields,

$$\int_\Gamma (w - V[w])\mu \, ds = \int_\Gamma u^o \mu \, ds \tag{2.14}$$

We combine equations (2.13) and (2.14) as follows. Define a Hilbert space H by,

$$H = \{(w,\lambda) : w \in H_1(B), \lambda \in H_{-1/2}(\Gamma)\} \tag{2.15}$$

with the norm,

$$\| \{w,\lambda\} \|^2_H = \| w \|^2_{1,B} + \| \lambda \|^2_{-1/2,\Gamma} \tag{2.16}$$

Define the bilinear form $a(\{w,\lambda\}, \{\phi,\mu\})$ on $H \times H$ by,

$$\begin{aligned} a(\{w,\lambda\},\{\phi,\mu\}) &= - \int_B(\nabla w \cdot \nabla \phi + qw\phi)dx + \int_\Gamma \lambda\phi ds \\ &\quad + \int_\Gamma(w - V[w])\mu ds \end{aligned} \tag{2.17}$$

and the functional F by,

$$F(\{\phi,\mu\}) = \int_\Gamma u^o \mu ds \tag{2.18}$$

Then (2.13) and (2.14) imply that,

$$a(\{w,\lambda\},\{\phi,\mu\}) = F(\{\phi,\mu\}) \quad \text{for all} \quad \{\phi,\mu\} \in H \tag{V_{II}}$$

This is our second variational procedure. If $\{w,\lambda\}$ is a solution, and w has enough smoothness, then w will solve (P_{II}) and λ will be its normal derivative. Boundary conditions are natural and it can be shown that numerical implementation with finite elements yields sparse matrices, see [4] for a related calculation.

3. SOME REMARKS ON VALIDITY

In this section we want to give some indications of what must be done to show that our two variational procedures are well posed. For the first method the result is easy.

THEOREM I. Problem (V_I) has a unique solution.

Proof: We will establish the following inequality for the bilinear form $a(w,\phi)$ of (2.2):

$$-a(w,w) \geq m\| w \|_H^2, \qquad m > 0 \tag{3.1}$$

Once (3.1) is verified the conclusion of the theorem follows from the Lax-Milgram lemma. Since $q > 0$ in B the inequality (3.1) is an immediate consequence of the following result.

LEMMA: For any $\phi \in H$, $\int_{\Gamma_\infty} T[\phi]\phi ds \leq 0$

Proof of the Lemma. Choose a circle C_R containing Γ_∞ in its interior and let D_R denote the region between Γ_∞ and C_R. Let v be the solution of (P_∞). Then by Green's theorem we have,

$$\int_{D_R} (v\Delta v + |\nabla v|^2)dx = \int_{D_R} (|\nabla v|^2 + v^2)dx = \int_{\partial D_R} v \frac{\partial v}{\partial n} ds$$

$$= - \int_{\Gamma_\infty} \phi T[\phi]ds + \int_{C_R} v \frac{\partial v}{\partial n} ds$$

Solutions of (P_∞) actually vanish exponentially as $|s| \to \infty$. Hence the second term on the right tends to zero as R tends to infinity and the conclusion of the lemma follows.

THEOREM II. Problem (P_{II}) has a unique solution.

The proof of Theorem II is much more complex and we will only sketch it. The source of difficulty is that the bilinear form $a(\{w,\lambda\},\{\phi,\mu\})$ in (2.17) is not negative definite so that it does not satisfy an inequality of the form (3.1). Thus one cannot use the Lax-Milgram lemma. There is, however, a substitute. It follows from results in [1] that a sufficient condition that (V_{II}) have a solution is that a satisfy the inequality,

$$\sup_{\{\phi,\mu\}\in H} \frac{|a(\{w,\lambda\},\{\phi,\mu\})|}{\|\{\phi,\mu\}\|_H} \geq m\|\{w,\lambda\}\|_H, \qquad m > 0 \tag{3.2}$$

We outline a procedure for establishing the inequality (3.2). Given $\{w,\lambda\} \in H$ we seek a $\{\phi,\mu\}$ so that (3.2) holds by taking $\phi = w + \psi$. From (2.17) we have,

$$\begin{aligned} a(\{w,\lambda\},\{w+\psi,\mu\}) &= -\int_B (|\nabla w|^2 + q|w|^2)dx \\ &\quad - \int_B (\nabla w\cdot\nabla\psi + qw\psi - wK(\mu))dy + \int_\Gamma (\lambda w + \lambda\psi + w\mu)ds, \end{aligned} \tag{3.3}$$

where,

$$K[\mu](y) = \int_\Gamma G(x,y)(q(y) - 1)\mu(x)ds_x \tag{3.4}$$

Now we need yet another auxiliary problem. We first choose $\psi_o \in H_{1/2,\Gamma}$ such that,

$$\int_\Gamma \lambda\psi_o ds = -\|\lambda\|_{-1/2,\Gamma}\|\psi_o\|_{1/2,\Gamma} \tag{3.5}$$

Then we define $\{\psi,\mu\} \in H$ as a solution of the problem,

$$\begin{aligned} \Delta\psi - q\psi &= -K(\mu) \quad \text{in} \quad B \\ \psi &= \psi_o \quad \text{on} \quad \Gamma \\ \psi_n - \mu &= \lambda \quad \text{on} \quad \Gamma \end{aligned} \tag{P^*_{II}}$$

Suppose $\{\psi,\mu\}$ solves (P^*_{II}). Then integration by parts yields,

$$-\int_B (\nabla w\cdot\nabla\psi + qw\psi - wK(\mu)) + \int_\Gamma (w\mu + w\lambda)ds = 0 \tag{3.6}$$

It is possible to show, by means of integral equation arguments, that (P^*_{II}) has a solution and that for this solution one has the a priori bounds,

$$\|\psi\|_{1,B} \leq K\|\lambda\|_{-1/2,B} \qquad \|\mu\|_{-1/2,\Gamma} \leq K\|\lambda\|_{-1/2,\Gamma} \tag{3.7}$$

This argument is quite technical and will be presented elsewhere. It follows from (3.5) that

$$\|\lambda\|_{-1/2,B} = \|\psi_o\|_{1/2,B}$$

and hence (3.5) yields,

$$\int_\Gamma \lambda\psi ds \leq -K' \|\lambda\|^2_{-1/2,B} \tag{3.8}$$

Now we combine (3.3), (3.6) and (3.8) to obtain,

$$\begin{aligned} a(\{w,\lambda\},\{w+\psi,\mu\}) &= -\int_B (|\nabla w|^2 + qw^2)dx + \int_\Gamma \lambda\psi ds \\ &\leq -m'(\int_B (|\nabla w|^2 + w^2)dx + \|\lambda\|^2_{-1/2,B}) = -m'\|\{w,\lambda\}\|^2_H \end{aligned} \tag{3.9}$$

for some $m' > 0$. But we have, by (3.7),

$$\| \{w + \psi,\mu\} \|_H \le C(\| w \|_{1,B} + \| \psi \|_{1,B} + \| \mu \|_{-1/2,\Gamma})$$
$$\le C(\| w \|_{1,B} + 2K\| \lambda \|_{-1/2,\Gamma}) \le C'\| \{w,\lambda\} \|_H \tag{3.10}$$

If we set $\phi = w + \psi$ then (3.9) and (3.10) yield,

$$|a(\{w,\lambda\},\{\phi,\mu\})| \ge \frac{m'}{C'} \| \{w,\lambda\} \|_H \| \{\phi,\mu\} \|_H$$

and this establishes (3.2).

It is interesting to note that the problem (P^*_{II}) is a kind of adjoint to (P_{II}). For the corresponding homogeneous problems the adjointness becomes exact. Corresponding to (P_{II}) we define L by,

$$D_L = \{w : w \in C^{(2)}(B) \cap C^{(1)}(\bar{B}), \qquad w = V[w] \quad \text{on} \quad \Gamma\}$$

$$Lw = \Delta w - qw$$

Corresponding to (P^*_{II}) we define L^* by

$$D_{L^*} = \{\psi : \psi \in C^{(2)}(B) \cap C^{(1)}(\bar{B}), \qquad \psi = 0 \quad \text{on} \quad \Gamma\}$$

$$L^*\psi = \Delta\psi - q\psi + K[\frac{\partial\psi}{\partial n}]$$

Then we have, for $w \in D_L$ and $\psi \in D_{L^*}$

$$\int_B Lw\psi dy = \int_B (\Delta w - qw)\psi dy = \int_\Gamma (w_n\psi - w\psi_n)ds$$
$$+ \int_B (\Delta\psi - q\psi)wdy = - \int_\Gamma V[w]\psi_n ds + \int_B (\Delta\psi - q\psi)wdy$$
$$= \int_B (\Delta\psi - q\psi - K[\psi_n])wdy = \int_B L^*\psi wdy$$

Notice the contrast between the two problems (P_I) and (P_{II}). One can show by a very similar argument to that in the proof of the lemma that the operator T satisfies

$$\int_\Gamma T[\phi]\psi ds = \int_\Gamma \phi T[\psi]ds$$

Thus the adjoint homogeneous problem for (P_I) is the same as the homogeneous problem. For (P_{II}), however, the adjoint of the nonlocal boundary condition problem is a problem with a local boundary condition but a nonlocal equation.

ACKNOWLEDGMENTS

This work was supported by National Science Foundation Grant MCS 77-01449.

REFERENCES

1. Babuska, I. and A. K. Aziz, *Survey Lectures on the Mathematical Foundations of the Finite Element Method*, Academic Press, New York, 1972.

2. Barrar, R. B. and C. L. Dolph, "On a three-dimensional transmission problem of electromagnetic theory", *Journal of Rat. Mech. and Anal.* 3 (1954), 725-743.

3. Duffin, R. J. and J. H. McWhirter, "An integral equation formulation of Maxwell's equation", *Journ. of the Franklin Institute* 298 (5,6) (1974), 385-394.

4. Fix, G. J., "Hybrid finite element methods," *Numerical Simulation of Fluid Motion*, edited by John Noye, North Holland Publishing Co., New York, 1978.

5. Hsiao, G. -C. and R. C. MacCamy, "Solutions of boundary value problems by integral equations of the first kind", *SIAM Review* 15 (4) (1973).

6. Hsiao, G. -C. and W. L. Wendland, "A finite element for some integral equations of the first kind", *Journ. of Math. Anal. and Appl.* 58 (3) (1977).

7. MacCamy, R. C. and S. P. Marin, "A finite element method for exterior interface problems," *International Journal of Mathematics and Mathematical Sciences*, Vol. 3, (2) (1980), 311-350.

8. MacCamy, R. C. and J. H. McWhiter, "Eddy currents in thin sheets", Westinghouse Research Laboratory Technical Report.

NONLOCAL BOUNDARY PROBLEMS FOR FUNCTIONAL PARTIAL DIFFERENTIAL EQUATIONS

Moshe Marcus*
Victor J. Mizel

Department of Mathematics
Carnegie-Mellon University
Pittsburgh, Pennsylvania

1. PRELIMINARIES

Although it was realized by the early eighteenth century that for some physical systems the prediction of future evolution requires knowledge regarding *past* as well as present states, it is only during the last twenty years or so that mathematicians have begun to analyze such phenomena in a broadly spread effort. It was emphasized by Krasovskii in the early 1960's that progress in the study of such equations of evolution should ensue from a formulation featuring states which are *functions* incorporating past and present data. Nowadays this function space approach is quite standard, and the term "history space" is often used to underline the role of the function spaces encountered in such studies.

There are two main types of function spaces which have been employed in these evolution equation investigations:

(1) Spaces of continuous functions on a compact interval, which in the U.S. have been thoroughly explored by Hale and his colleagues (cf. [11] and the references cited therein).

**Current affiliation:* Department of Mathematics, Technion-Israel Institute of Technology, Haifa, Israel

(2) Appropriately weighted Lebesgue spaces on a real interval, which were introduced, for applications to continuum mechanics, by Coleman and Mizel [4,5] in the case of the half axis (cf. [6] for a related class of spaces which had been introduced into continuum mechanics previously)--until recently such spaces were mainly exploited by researchers having strong interests in continuum mechanics [18,7,8,16]. The framework of Lebesgue spaces on a bounded interval was also employed by Borosovic and Turbabin [1] for certain linear problems on a bounded interval (see also [14]).

Despite the successes of numerous investigations involving this history space framework, there has been a less well publicized minority view, expressed by G. S. Jones [13] and others, which stresses the importance of formulating problems directly in terms of sets of trajectories within the *elementary* state space associated with the phenomenon, since the formulation in terms of states belonging to a fixed history space can be too narrow. This latter view has been vindicated by the recent elegant results obtained for linear problems "ignorant of the future" by Coffman and Schäffer [2,3], where such a trajectory formulation was indeed crucial to the analysis.

It will be seen that the present work which is in essence a status report on an ongoing collaboration in these matters which began in 1977, has, despite its primary emphasis on functional *partial* differential equation ("FPDE") problems, interesting implications for functional differential equations ("FDEs") and neutral functional differential equations ("NFDEs"), as well. In particular, our framework serves to incorporate, in a Lebesgue space context, practically all operators previously employed in FDE and NFDE problems, some of which would seem to make sense only in a space of continuous functions.

In order to motivate the framework which will be erected in Section 2 for FPDE, we proceed now to describe the form that the framework takes and the results that it provides for a broad class of FDE and NFDE problems.

Let $\rho > 0$ a.e. be a "pseudo-influence function" on $\mathbb{R}^- = (-\infty,0]$ in the sense that ρ is a Borel function satisfying

$$\text{(i)} \quad \operatorname*{ess\,sup}_{\theta \in \mathbb{R}^-} \frac{\rho(\theta-\sigma)}{\rho(\theta)} = \overline{K}(\sigma) < \infty \qquad \text{for each} \quad \sigma > 0 \tag{1.1a}$$

$$\text{(ii)} \quad \overline{K} : \mathbb{R}^+ \to \mathbb{R}^+ \qquad \text{is bounded on each interval } [0,A]$$

It then follows [4,16] that the Banach spaces $L_p(\rho) := L_p(\mathbb{R}^-,\rho;R_N)$, $1 \leq p < \infty$, [N is fixed throughout] all possess a strongly continuous semigroup of ("left") translation operators $\{T^\sigma\}_{\sigma \geq 0}$ defined on measurable functions $f \in L_p(\rho)$ by

$$(T^\sigma f)(\theta) = \begin{cases} f(\theta + \sigma) & \theta \leq -\sigma \\ 0 & -\sigma < \theta \leq 0 \end{cases} \tag{1.1b}$$

Given any function f defined on an interval $(-\infty,A]$ and any $t \in (-\infty,A]$, f_t will denote the function on $\mathbb{R}^-$ ("history") defined by:

$$f_t(\theta) = f(t + \theta), \qquad \theta \in \mathbb{R}^- \tag{1.1c}$$

A type of functional differential equation problem frequently encountered [5,7] on such spaces can be formulated as

$$\begin{aligned} &\dot{x}(t) = A(t,x(t),x_t), \qquad 0 \leq t \leq T, \\ &x_o(\cdot) = \phi \in L_p(\rho), \qquad x_o(0) = a \end{aligned} \tag{1.2}$$

where $A : [0,T] \times R_N \times L_p(\rho) \to R_N$ is continuous and bounded in its last two arguments for a.e. $t \in [0,T]$, is an integrable function in its first argument for each $(b,w) \in R_N \times L_p(\rho)$ and is such that for each $M > 0$ the functions $\{|A(\cdot,b,w)| : \|w\|_{L_p(\rho)} \leq M, |b| \leq M\}$ are dominated by a function $\Psi_M \in L^1[0,T]$. What we wish to emphasize here is that the functional A in (1.2) generates an *operator* $\mathcal{A}$ acting on the class $\lambda^p_{0,T}(\rho)$ of those R_N-valued Borel functions y on $(-\infty,T]$ which are continuous on $[0,T]$ and are such that y_o is in $L_p(\rho)$. Namely, (abusing notation a bit) let

$$\mathcal{A}(y)(t) = A(t,y(t),y_t), \qquad t \in [0,T], \qquad y \in \lambda^p_{0,T}(\rho) \tag{1.3}$$

Then our assumptions on A imply that $\mathcal{A}$ maps each such "process" y into an integrable function on [0,T] and, furthermore, that elements y,z which are almost everywhere equal on $(-\infty,T]$ are mapped into the same equivalence class of $L^1[0,T]$. Moreover, questions of existence and uniqueness for (1.2) can be very efficiently described in terms of properties of the operator $\mathcal{A} : \lambda^p_{0,T}(\rho) \to L^1[0,T]$.

In line with the viewpoint expressed in the preceding paragraph, we proceed to discuss a far broader class of functional differential equations than (1.2). Let $\Lambda^p_{0,T}(\rho) \supset \lambda^p_{0,T}(\rho)$ denote the class of processes y on $(-\infty,T]$ which belong to L^p on $[0,T]$ and satisfy $y_0 \in L_p(\rho)$. Consider the equation

$$\dot{x}(t) = G(x)(t), \qquad 0 \le t \le T, \qquad (1.4)$$
$$x_o(\cdot) = \phi \in L_p(\rho), \qquad x_o(0) = \underset{\sim}{a}$$

where $G : \lambda^p_{0,T}(\rho) \to L^1[0,T]$ or $G : \Lambda^p_{0,T}(\rho) \to L^1[0,T]$ denotes *any* operator on the relevant class of processes which maps each y to an integrable function on [0,T] (in such a manner as to preserve equivalence classes) and which satisfies the following "nonanticipativity" requirement (cf. the use of this notion by Coffman and Schäffer [2]):*

(NA) $$y\Big|_{(-\infty,\tau]} \doteq z\Big|_{(-\infty,\tau]} \Rightarrow G(y)\Big|_{[0,\tau]} \doteq G(z)\Big|_{[0,\tau]}$$

for $y,z \in \lambda^p_{0,T}(\rho)$ [resp. $y,z \in \Lambda^p_{0,T}(\rho)$] and for any (1.5)

$\tau \in [0,T]$

It is easy to verify that the operator A determined by the right side of (1.2) satisfies these conditions on $\lambda^p_{[0,T]}(\rho)$. However examples of a different sort are easily brought to mind. For instance, one can consider Hammerstein type operators of the form (cf. [16,17] for their use in an integral equation context):

$$G_1(y)(t) = \int_{-\infty}^{0} a(s)g(t + s, y(t + s))ds \qquad (1.6)$$

where $u \to g(\cdot,u(\cdot))$ is a composition operator taking $\Lambda^p_{0,T}(\rho)$ into $\Lambda^1_{0,T}(\rho)$ and the function a has the form $a(s) = \rho(s)b(s)$, with $b \in L^1(\mathbb{R}^-)$; alternatively one can consider operators having the form

$$G_2(y)(t) = g(t, \int_{-\infty}^{0} y_t(s)d\mu(s)) \qquad (1.7)$$

where μ is a finite Borel measure having compact support and $g : [0,T] \times R_N \to R_N$ is a Nemytskii operator carrying $L^p[0,T]$ into $L^1[0,T]$ [15], thereby including such familiar operators as

$$G(y)(t) = g(t,y(t),y(t - \theta_o),\dots,y(t - \theta_k))$$

Both G_1 and G_2 can be easily verified to fit into the class occurring in (1.4).

We now state certain existence-uniqueness results for (1.4) which arise as corollaries of the results to be obtained for FPDE in Section 2. It will be convenient to introduce the following notation. For $u : [0,T] \to R$ and $\phi : \mathbb{R}^- \to R$ set

---

* "$\doteq$" signifies "equality almost everywhere."

$$(u\square\phi)_t(\theta) = \begin{cases} u(t+\theta) & 0 \le t+\theta \le T \\ \phi(t+\theta) & t+\theta < 0, \quad t \in [0,T] \end{cases} \tag{1.8}$$

Then put

$$X^p_{0,T}(\phi) = \{x : x = u\square\phi, \quad u \in L_p(0,T)\}$$

$$X^c_{0,T}(\phi) = \{x : x = u\square\phi, \quad u \in C[0,T]\}$$

THEOREM 1.1. Suppose that the nonanticipative operator $G : \Lambda^p_{0,T}(\rho) \to L^1[0,T]$ is continuous and bounded relative to the following norm on $\Lambda^p_{0,T}(\rho)$:

$$|||x|||_{p;\rho} = \sup_{\tau\in[0,T]} \{\| x_\tau \|_{L_p(\rho)}\}, \qquad x \in \Lambda^p_{0,T}(\rho) \tag{1.9}$$

Then there exists for each $\phi \in L^p(\rho)$ and $\underset{\sim}{a} \in R_N$ a $t_0 > 0$ and a function x on $(-\infty,t_0]$, absolutely continuous on $[0,t_0]$, which satisfies (1.4). If in addition G satisfies the growth condition

$$\| G(x) \|_{L_1[0,T]} \le \text{const}(1 + |||x|||_{p;\rho}), \qquad x \in \Lambda^p_{0,T}(\rho) \tag{1.10}$$

then there exists a solution of (1.4) on $(-\infty,T]$ for each such pair ϕ, $\underset{\sim}{a}$.

Furthermore the solution discussed above will be unique provided that G satisfies the following Lipschitz condition

$$\| G(x) - G(x^*) \|_{L^1(0,T)} \le \text{const}|||x - x^*|||_{p;\rho}, \qquad x,x^* \in \Lambda^p_{0,T}(\rho) \tag{1.11}$$

and this solution will depend continuously on the initial data ϕ, $\underset{\sim}{a}$.

Somewhat related problems were considered in [14,9,19,12].

Naturally the hypotheses above can be given in a more local way. The following result involves such a formulation.

THEOREM 1.1.* Given $\phi \in L_p(\rho)$, suppose that G is a nonanticipative operator from $X^p_{0,T}(\phi)$ into $L^1[0,T]$ such that the mapping $\tilde{G} : L^p[0,T] \to L^1[0,T]$ which is defined by

$$\tilde{G}(u) := G(u\square\phi) \tag{1.12}$$

is continuous and bounded. Then for each $\underset{\sim}{a} \in R_N$ there exists a local solution x of (1.4) defined on some interval $(-\infty,t_0]$. If in addition $\tilde{G}$ satisfies the growth condition

$$\| \tilde{G}(u) \|_{L_1(0,T)} \leq \text{const}(1 + \| u \|_{L^p(0,T)})$$

then there exists a solution on $(-\infty,T]$ for each $\tilde{a} \in R_N$.

If $\tilde{G}$ satisfies the Lipschitz condition

$$\| \tilde{G}(u) - \tilde{G}(u^*) \|_{L_1(0,T)} \leq \text{const} \| u - u^* \|_{L_p(0,T)}$$

then the solution discussed above will be unique.

There are analogous existence-uniqueness results available when G is an operator on $\lambda^p_{0,T}(\rho)$ or $X^c_{0,T}(\phi)$, respectively. For brevity we limit ourselves here to stating the analogue of the "local" formulation given in Theorem 1.1*.

THEOREM 1.2.* Given $\phi \in L_p(\rho)$, suppose that G is a nonanticipative operator from $X^c_{0,T}(\rho)$ into $L^1[0,T]$ such that the mapping $\tilde{G} : C[0,T] \to L^1[0,T]$ which is defined by (1.12) is continuous and furthermore is such that, for each $M > 0$ the functions

$$\{\tilde{G}(u) : \| u \|_{L^\infty(0,T)} \leq M\}$$

are dominated by a function $\Psi_M \in L^1[0,T]$. Then for each $\tilde{a} \in R_N$ there exists a local solution x of (1.4) defined on some interval $(-\infty,t_0]$.

If $\tilde{G}$ satisfies the Lipschitz type condition

$$\| \tilde{G}(u) - \tilde{G}(u^*) \|_{L_1[0,t]} \leq \text{const} \int_0^t g(s) \| u - u^* \|_{L_\infty(0,s)} ds,$$

$$t \in [0,T]$$

for some $g \in L^1[0,T]$, then the solution discussed above will be unique.

REMARKS: (1) The hypotheses (1.1a) on ρ play no role above except in Theorem 1.1 where they ensure the existence of a solution of (1.4) for each "initial segment" $\phi \in L_p(\rho)$.

(2) It is easy to verify that the operator G associated with the right side of (1.2) satisfies the hypothesis of Theorem 1.2*, so that the latter result yields, in particular, a local existence theorem for such FDEs, as well as for FDEs with right side of Hammerstein type as in (1.6), or of the type given in (1.7).

In addition to furnishing an existence-uniqueness theory for the broad class of functional differential equations (1.4), the results to be given on FPDE also yield, as noted earlier, existence-uniqueness results for a wide class of "neutral" functional differential equations, including those of the form

$$\begin{aligned} \dot{x}(t) &= A(t,x(t),x_t) + \frac{d}{dt}(B(t,x_t)), \\ x_0(\cdot) &= \phi \in L_p(\rho), \qquad x_0(0) - B(0,x_0) = \underset{\sim}{a} \end{aligned} \tag{1.13}$$

We end this section by stating one existence-uniqueness result of this type for the equation

$$\begin{aligned} \dot{x}(t) &= G(x)(t) + \frac{d}{dt} H(x)(t), \qquad 0 \le t \le T \\ x_0(\cdot) &= \phi \in L_p(\rho), \qquad x_0(0) - H(x_0)(0) = \underset{\sim}{a} \end{aligned} \tag{1.14}$$

THEOREM 1.3. Suppose that G and H are nonanticipative operators such that $G : \Lambda^p_{0,T}(\rho) \to L^1[0,T]$, $H : \Lambda^p_{0,T}(\rho) \to L^p[0,T]$, with G continuous and bounded relative to the norm $|||\ |||_{p;\rho}$ on $\Lambda^p_{0,T}(\rho)$ and with H satisfying the Lipschitz type condition

$$\begin{aligned} \| H(x) - H(x^*) \|_{L_p(0,t)} &\le c(t) \sup_{\tau\in[0,t]} \{ \| x_\tau - x^*_\tau \|_{L_p(\rho)} \} \\ \text{for all} \quad t \in [0,T], &\qquad x,x^* \in \Lambda^p_{0,T}(\rho) \end{aligned} \tag{1.15}$$

with $c(t) \to 0$ as $t \to 0$. Then there exists for each $\phi \in L_p(\rho)$ and $\underset{\sim}{a} \in R_N$ a positive number t_0 ($t_0 \le T$) such that (1.14) possesses a solution x on $(-\infty,t_0]$.

This result includes operators H having the form (1.7), where g provides a Lipschitz mapping from $L^p[0,T]$ to $L^p[0,T]$ and μ has no atom at the origin.

2. FUNCTIONAL PARTIAL DIFFERENTIAL EQUATIONS

In this section we formulate the functional partial differential equation ("FPDE") problems to be considered, and we introduce certain function spaces which seem to be particularly well suited to such problems. Consider the following system of FPDEs and boundary FDEs on the semi-infinite strip $Q = Q_T = \mathbb{R}^- \times [0,T]$ in R_2:

$$\begin{cases} \dfrac{\partial x}{\partial t} - E(t,0)\,\dfrac{\partial x}{\partial \theta} = \mathcal{F}(\underset{\sim}{x}) \quad \text{in } Q & (2.1a) \\ \dfrac{\partial x}{\partial t}(t,0) = \mathcal{G}(\underset{\sim}{x}) + \dfrac{d}{dt}\mathcal{H}(\underset{\sim}{x}) \quad \text{in } [0,T] & (2.1b) \\ \underset{\sim}{x}(0,\cdot) = \phi \in L_p(\rho), \quad \underset{\sim}{x}(0,0) - \mathcal{H}(\underset{\sim}{x})(0) = \underset{\sim}{a} \in R_N & (2.1c) \end{cases}$$

where $\underset{\sim}{x}$ is an R_N-valued function and E is a diagonal $N \times N$ matrix function

$$E(t,\theta) = \begin{pmatrix} \lambda_1(t,\theta) & & 0 \\ & \ddots & \\ 0 & & \lambda_N(t,\theta) \end{pmatrix} \tag{2.2}$$

on Q, while $\mathcal{F}$, $\mathcal{G}$, $\mathcal{H}$ are operators on a function space $\mathcal{M}_p^{\rho,\tilde{\rho}}$, $1 \leq p < \infty$, which we now define.

DEFINITION 2.1. Given any pseudo-influence function ρ and $p,q \in [1,\infty)$, set $Q_\tau = \mathbb{R}^- \times [0,\tau]$ for each $\tau \in [0,T]$ and denote

$$\begin{aligned} &L^\rho_{p,q}(Q_\tau) := L_q([0,\tau];L_p(\rho)), \qquad L^\rho_{p,q} := L^\rho_{p,q}(Q_T) \\ &\| f \|_{L^\rho_{p,q}(Q_\tau)} := \left(\textstyle\int_0^\tau (\| f(t,\cdot) \|_{L_{p(\rho)}})^q dt\right)^{1/q}, \\ &\| f \|_{L^\rho_{p,q}} := \| f \|_{L^\rho_{p,q}(Q_T)} \end{aligned} \tag{2.3}$$

In addition, equip the following sets of functions with the norms indicated:

$$\begin{aligned} &A^\rho_p(Q_\tau) := C([0,T];\, L_p(\rho)), \qquad A^\rho_p := A^\rho_p(Q_T) \\ &\| f \|_{A^\rho_p(Q_\tau)} := \sup_{t\in[0,\tau]} \{\| f(t,\cdot) \|_{L_p(\rho)}\}, \\ &\| f \|_{A^\rho_p} := \| f \|_{A^\rho_p(Q_T)} \end{aligned} \tag{2.4}$$

$$\begin{aligned} &B^\rho_p := B^\rho_p(Q_T), \qquad B^\rho_p(Q_\tau) := \{f \in C(\mathbb{R}^-;L_p[0,\tau]) : \\ &\operatorname*{ess\,sup}_{\theta\in\mathbb{R}^-}(\rho(\theta)\| f(\cdot,\theta) \|_{L_p(0,\tau)}) < \infty\}, \\ &\| f \|_{B^\rho_p(Q_\tau)} := \operatorname*{ess\,sup}_{\theta\in\mathbb{R}^-}(\rho(\theta)\| f(\cdot,\theta) \|_{L_p(0,\tau)}), \\ &\| f \|_{B^\rho_p} := \| f \|_{B^\rho_p(Q_T)} \end{aligned} \tag{2.5}$$

Finally, given any pair of pseudo-influence functions $\rho,\tilde{\rho}$, set

$$\begin{aligned} &M_p^{\rho,\tilde{\rho}}(Q_\tau) := A_p^{\rho}(Q_\tau) \cap B_p^{\tilde{\rho}}(Q_\tau) \\ &\| f \|_{M_p^{\rho,\tilde{\rho}}(Q_\tau)} := \| f \|_{A_p^{\rho}(Q_\tau)} + \| f \|_{B_p^{\tilde{\rho}}(Q_\tau)}, \qquad 1 \leq p < \infty \end{aligned} \tag{2.6}$$

Clearly all these normed spaces are complete.

The hypotheses to be required of the continuous operators F, G, H occuring in (2.1) can now be described as follows:

$$\begin{aligned} &F: M_p^{\rho,\tilde{\rho}} \to L_{p,1}^{\rho}, \qquad G: M_p^{\rho,\tilde{\rho}} \to L_1[0,T], \\ &H: M_p^{\rho,\tilde{\rho}} \to L_p[0,T], \qquad p \in [1,\infty) \end{aligned} \tag{2.7}$$

In addition we require that F, G, H are nonanticipative in a sense analogous to (1.5):*

$$y\Big|_{Q_\tau} \overset{\text{a}}{=} z\Big|_{Q_\tau} \Rightarrow \begin{cases} F(y)\Big|_{Q_\tau} \overset{\text{a}}{=} F(z)\Big|_{Q_\tau} \\ G(y)\Big|_{[0,\tau]} \doteq G(z)\Big|_{[0,\tau]} \\ H(y)\Big|_{[0,\tau]} \doteq H(z)\Big|_{[0,\tau]} \end{cases} \tag{2.8}$$

$$\text{for } y,z \in M_p^{\rho,\tilde{\rho}} \text{ and each } \tau \in [0,T]$$

It should be noted that (2.1) is a broad enough formula to include a wide variety of applications. For example, (2.1) includes an interesting model in population dynamics due to Gurtin and MacCamy [10], where E is the identity and $G \equiv 0$.

The simplest nontrivial examples of operators F, G, H are obtained by

$$\begin{aligned} &F(z)(t,\theta) = F(t,\theta;z(t,\theta),z(t,\cdot),z(\cdot,\theta)) \\ &G(z)(t) = G(t;z(t,0),z(t,\cdot)), \qquad H(z)(t) = H(t;z(t,\cdot)) \end{aligned} \tag{2.9}$$

where $F : Q \times R_N \times L_p(\rho) \times L_p[0,T] \to R_N$ is continuous and bounded in its last three arguments for a.e. $(t,\theta) \in Q$, is an $L_{p,1}^{\rho}$ function in (t,θ) for each $(\underset{\sim}{b},\underset{\sim}{w},\underset{\sim}{v}) \in R_N \times L_p(\rho) \times L_p[0,T]$ and is such that for each $M > 0$ the functions

$$\begin{aligned} &\{F(\cdot,\cdot;\underset{\sim}{b},\underset{\sim}{w},\underset{\sim}{v}) : \| \underset{\sim}{w} \|_{L_p(\rho)} \leq M, \\ &\qquad \operatorname*{ess\,sup}_{\theta \in R^-} \{(\tilde{\rho}(\theta) \| v \|_{L_p(0,T)}) \leq M\} \end{aligned} \tag{2.10}$$

---

*$\overset{\text{a}}{=}$ denotes "equality almost everywhere" (relative to planar Lebesgue measure).

are dominated by a function $\xi_M + C_M|\underset{\sim}{b}|^p$, $\xi_M \in L^\rho_{p,1}$; and where $G : [0,T] \times R_N \times L_p(\rho) \to R_N$ and $H : [0,T] \times L_p(\rho) \to R_N$ are continuous and bounded in their other arguments for a.e. $t \in [0,T]$, are respectively integrable and p-summable in t for each $(\underset{\sim}{b},\underset{\sim}{w}) \in R_N \times L_p(\rho)$ [for each $\underset{\sim}{w} \in L_p(\rho)$] and are such that for each $M > 0$ the functions

$$\{G(\cdot;\cdot,\underset{\sim}{w}) : \|\underset{\sim}{w}\|_{L_p(\rho)} \leq M\}, \qquad \{H(\cdot;\underset{\sim}{w}) : \|\underset{\sim}{w}\|_{L_p(\rho)} \leq M\} \tag{2.11}$$

are dominated respectively by expressions of the form

$$\begin{aligned} &\Psi_M(t) + c_M|\underset{\sim}{b}|^p, \qquad \zeta_M(t) \qquad \text{with} \\ &\qquad \Psi_M \in L_1[0,T], \qquad \zeta_M \in L_p[0,T] \end{aligned} \tag{2.12}$$

On the other hand the requirements (2.7) and (2.8) are obviously also satisfied by a wide variety of other types of operators as well, as can be illustrated along the lines of (1.6) and (1.7).

Concerning the matrix function E in (2.1) we require that the functions λ_k in (2.2) belong to $C^1(Q)$ and satisfy

$$\frac{1}{C_0} \leq \lambda_k \leq C_0, \qquad \left|\frac{\partial\lambda_k}{\partial\theta}\right| \leq C_0 \qquad k = 1,\dots,N \tag{2.13}$$

for some $C_0 > 0$. These requirements on E can be considerably weakened. We shall denote the characteristics associated with E by γ_k where $\gamma_k(\cdot;\bar{t},\bar{\theta})$, $k = 1,\dots,N$, are the characteristic curves of (2.1_a) through the point $(\bar{t},\bar{\theta}) \in Q$. That is,

$$\frac{d\gamma_k}{dt} = -\lambda_k(t,\gamma_k), \qquad \gamma_k(\bar{t};\bar{t},\bar{\theta}) = \bar{\theta} \tag{2.14}$$

Then (2.13) ensures that (2.14) has a unique solution in $C^1[\beta_k(\bar{t},\bar{\theta}),T]$, where $\beta_k(\bar{t},\bar{\theta})$ denotes the t-coordinate of the point where the solution of (2.14) intersects ∂Q below $t = \bar{t}$. For each k we decompose $Q = Q_T$ into two subregions according to

$$Q_{T,k} = \{(\bar{t},\bar{\theta}) : \beta_k(\bar{t},\bar{\theta}) = 0\}, \qquad Q'_{T,k} = \{(\bar{t},\bar{\theta}) : \beta_k(\bar{t},\bar{\theta}) > 0\} \tag{2.15}$$

and we denote by $\theta = \alpha_k(\bar{t},\bar{\theta})$ the function defined on $Q_{T,k}$ by

$$\alpha_k(\bar{t},\bar{\theta}) = \gamma_k(0;\bar{t},\bar{\theta}) \tag{2.16}$$

We now define what is meant by a (weak) solution of (2.1). Integration of (2.1a), along its characteristic curves formally leads to the system of "characteristic" equations:

$$x_k(t,\theta) = \begin{cases} \phi_k(\alpha_k(t,\theta)) + \int_0^t F_k(\underset{\sim}{x})(s,\gamma_k(s;t,\theta))ds \\ \qquad (t,\theta) \in Q_{T,k} \\ x_k(\beta_k(t,\theta),0) + \int_{\beta_k(t,\theta)}^t F_k(\underset{\sim}{x})(s,\gamma_k(s;t,\theta))ds \\ \qquad (t,\theta) \in Q'_{T,k} \end{cases} \tag{2.17}$$

where $F = (F_1,\ldots,F_N)$, $\phi = (\phi_1,\ldots,\phi_N)$. Similarly, integration of (2.1b) formally leads to

$$x(t,0) = \int_0^t G(\underset{\sim}{x})(s)ds + H(\underset{\sim}{x})(t) + a, \qquad 0 \le t \le T \tag{2.18}$$

By definition a weak solution of (2.1) is any Borel function $\underset{\sim}{x} \in M_p^{\rho,\tilde{\rho}}$ which satisfies (2.17) and (2.18) except for a set of points $(t,\theta) \in Q$ whose intersection with *every* line t = const and with *every* line θ = const is a Lebesgue null set.

Our main results are the following existence-uniqueness theorems for the problem (2.1).

THEOREM 2.1. Given $p,q \in [1,\infty)$ let $\rho,\tilde{\rho}$ be pseudo-influence functions satisfying

$$\tilde{\rho}(\theta) \le \rho(\theta)^{1/p} \qquad \text{a.e. on } \mathbb{R}^- \tag{2.19}$$

Suppose that E satisfies (2.13) and that the nonanticipative operators F, G, H satisfy

(a) $G : M_p^{\rho,\tilde{\rho}} \to L_1[0,T]$ is continuous and obeys the growth condition

$$\| G(\underset{\sim}{x}) \|_{L_1[0,T]} \le \text{const}(1 + \| \underset{\sim}{x} \|_{M_p^{\rho,\tilde{\rho}}}), \qquad \underset{\sim}{x} \in M_p^{\rho,\tilde{\rho}};$$

(b) $H : M_p^{\rho,\tilde{\rho}} \to L_p[0,T]$ is continuous and satisfies the Lipschitz type condition

$$\| H(\underset{\sim}{x}) - H(\underset{\sim}{x}^*) \|_{L_p[0,\tau]} \le \left(\int_0^\tau g(s)(\| \underset{\sim}{x} - \underset{\sim}{x}^* \|_{M_p^{\rho,\tilde{\rho}}(Q_s)})^q ds\right)^{1/q},$$

for all $\tau \in [0,T]$

where g is a nonnegative function in $L_1[0,T]$;

(c) $F : M_p^{\rho,\tilde{\rho}} \to L_{p,1}^{\rho}$ is continuous and satisfies the Lipschitz type condition

$$\| F(x) - F(x^*) \|_{L_{p,1}^{\rho}(Q_\tau)} \leq (\int_0^\tau g(s)(\| x - x^* \|_{M_p^{\rho,\tilde{\rho}}(Q_s)})^q ds)^{1/q},$$

for all $\tau \in [0,T]$

with g as in (b).

Then there exists a weak solution of (2.1) on Q for each $\underset{\sim}{\phi} \in L_p(\rho)$, $\underset{\sim}{a} \in R_N$.

There is also available a version of this result involving local hypotheses.

THEOREM 2.2. Given $p \in [1,\infty)$ and two pseudo-influence functions $\rho, \tilde{\rho}$ satisfying (2.19), suppose that E satisfies (2.13). Given $\phi \in L_p(\rho)$, let $\hat{\underset{\sim}{\phi}} = (\hat{\phi}_1,\ldots,\hat{\phi}_N)$ denote the function in $M_p^{\rho,\tilde{\rho}}$ defined by

$$\hat{\phi}_k(t,\theta) = \begin{cases} \phi(\alpha_k(t,\theta)) & (t,\theta) \in Q_{T,k} \\ 0 & (t,\theta) \in Q'_{T,k} \end{cases}$$

Suppose that the operators F, G, H are defined and nonanticipative on some $M_p^{\rho,\tilde{\rho}}$-ball $B_R(\hat{\phi})$ centered at $\hat{\phi}$, and satisfy

(a*) $G : B_R(\hat{\phi}) \to L_1[0,T]$ is continuous and bounded;

(b*) $H : B_R(\hat{\phi}) \to L_p[0,T]$ satisfies the Lipschitz type condition

$$\| H(\underset{\sim}{x}) - H(\underset{\sim}{x}^*) \|_{L_p[0,\tau]} \leq c(\tau) \| \underset{\sim}{x} - \underset{\sim}{x}^* \|_{M_p^{\rho,\tilde{\rho}}(Q_\tau)}$$

for all $\tau \in [0,T]$

where $c(\tau) \downarrow 0$ as $\tau \to 0$;

(c*) $F : B_R(\hat{\phi}) \to L_{p,1}^{\rho}[0,T]$ satisfies the Lipschitz type condition

$$\| F(\underset{\sim}{x}) - F(\underset{\sim}{x}^*) \|_{L_{p,1}^{\rho}(Q_\tau)} \leq c(\tau) \| \underset{\sim}{x} - \underset{\sim}{x}^* \|_{M_p^{\rho,\tilde{\rho}}(Q_\tau)}$$

for all $\tau \in [0,T]$

with c as in (b*).

Then there exist positive numbers δ_0, t_0 ($t_0 \leq T$) such that (2.1) possesses a solution $\underset{\sim}{x}$ on Q_{t_0}. Moreover there exists a solution on the strip Q_{t_0} for each $\underset{\sim}{\psi} \in L_p(\rho)$, $\underset{\sim}{b} \in R_N$ satisfying

$$\| \psi - \phi \|_{L_p(\rho)} \leq \delta_0, \qquad |\underset{\sim}{b} - \underset{\sim}{a}| \leq \delta_0$$

The proofs of Theorems 2.1 and 2.2 make use of an interplay between the hyperbolic FPDE in (2.1) with *given* boundary data and the NFDE boundary expression in (2.1), and utilize a fixpoint theorem due to Darbo. The details of these arguments as well as a proof of several additional results for FPDE as well as for NFDE, FDE will be presented elsewhere.

In order to see the connection of (2.1) with the FDE (1.4) and NFDE (1.14) we need only take E to be the identity matrix and take $\underset{\sim}{F} \equiv 0$. We then associate to each $f \in \Lambda^p_{0,T}(\rho)$ a corresponding element $[f] \in M_p^{\rho,\tilde{\rho}}$ by

$$[f](t,\theta) = f(t + \theta) \qquad (t,\theta) \in Q$$

A similar identification applies to functions in $X^p_{[0,T]}(\phi)$ or $X^c_{[0,T]}(\phi)$. Under these identifications the relation between the three problems (2.1), (1.4), and (1.14) is readily seen.

ACKNOWLEDGMENT

Research partially supported by U.S. National Science Foundation Grant MCS-7703643-02.

REFERENCES

1. Borosovic, Jr. G. and A. S. Turbabin, "On the Cauchy problem for linear nonhomogeneous differential equations with retarded argument", *Soviet Math. Dokl.* 10 (1969), 401-405.

2. Coffman, C. V. and J. J. Schäffer, "Linear differential equations with delays: Existence, uniqueness, growth and compactness under natural Caratheodory conditions", *J. Diff. Eqns.* 16 (1974), 26-44.

3. Coffman, C. V. and J. J. Schäffer, "The structure of translation invariant memories", *J. Diff. Eqns.* 16 (1974), 428-459.

4. Coleman, B. D. and V. J. Mizel, "Norms and semigroups in the theory of fading memory", *Arch. Rat. Mech. Anal.* 23 (1966), 87-123.

5. Coleman, B. D. and V. J. Mizel, "On the stability of solutions of functional-differential equations", *Arch. Rat. Mech. Anal.* 30 (1968), 173-196.

6. Coleman, B. D. and W. Noll, "An approximation theorem for functionals, with applications to continuum mechanics", *Arch. Rat. Mech. Anal.* 6 (1960), 355-370.

7. Coleman, B. D. and D. R. Owen, "On the initial value problem for a class of functional-differential equations", *Arch. Rat. Mech. Anal.* 55 (1974), 275-299.

8. Dafermos, C., "Asymptotic stability in viscoelasticity", *Arch. Rat. Mech. Anal.* 37 (1970), 297-308.

9. Delfour, M. C. and S. K. Mitter, "Hereditary differential systems with constant delays. I. General case", *J. Diff. Eqns.* 12 (1972), 213-235.

10. Gurtin, M. E. and R. C. MacCamy, "Non-linear age-dependent population dynamics", *Arch. Rat. Mech. Anal.* 54 (1974), 281-300.

11. Hale, J. K., *Theory of Functional Differential Equations*, Springer-Verlag, New York, 1977.

12. Herdman, T. L. and J. A. Burns, "Functional differential equations with discontinuous right hand side," *Volterra Equations, Lecture Notes in Mathematics #737*, S. O. Londen and O. J. Staffens, Eds., Springer-Verlag, New York, 99-106, (1979).

13. Jones, G. S., "Hereditary sturcture in differential equations", *Math. Syst. Th.* 1 (1967), 263-278.

14. Kappel, F. and W. Schappacher, "Autonomous nonlinear functional differential equations and averaging approximations", *Nonlin. Anal. Th., Meth. Appl.* 2 (1978), 391-422.

15. Krasnosellskii, M. A., P. P. Zabreiko, E. I. Pustylnik and P. E. Sobolevskii, *Integral Operators in Spaces of Summable Functions*, Noordhoff, Leyden, 1976.

16. Leitman, M. J. and V. J. Mizel, "On fading memory spaces and hereditary integral equations", *Arch. Rat. Mech. Anal.* 55 (1974), 18-51.

17. Leitman, M. J. and V. J. Mizel, "Hereditary laws and nonlinear integral equations on the line", *Adv. in Math.* 22 (1976), 220-266.

18. MacCamy, R. C., "Exponential stability for a class of functional differential equations", *Arch. Rat. Mech. Anal.* 40 (1971), 120-138.

19. Schumacher, K., "Existence and continuous dependence for functional-differential equations with unbounded delay", *Arch. Rat. Mech. Anal.* 67 (1978), 315-335.

SINOSOIDAL INPUT-PERIODIC RESPONSE IN NONLINEAR DIFFERENTIAL EQUATIONS CONTAINING DISCONTINUOUS ELEMENTS

Richard K. Miller

Mathematics Department
Iowa State University
Ames, Iowa

Anthony N. Michel

Department of Electrical Engineering
Iowa State University
Ames, Iowa

1. INTRODUCTION

We shall study existence of periodic solutions of nonlinear systems when forced sinosoidally. These systems will be studied by using describing function techniques. We give specific computable conditions which guarantee that when the describing function method predicts a periodic response, then the nonlinear system has a periodic solution which is nearly equal to the predicted one. Our conditions are quite natural in the sense that they are precise versions of the usual criteria, i.e., the linear part of the system must be a good low pass filter and the describing function should provide a good approximation to the response of the nonlinear element for nearly sinosoidal inputs. We note that because we allow discontinuous nonlinearities it is necessary to work with a generalized notion of solution. Moreover it will not be possible to discuss the uniqueness of the periodic solution.

Many papers discuss various aspects of the theoretical justification of describing functions. The results of Bass [1], Bergen and Frank [2] and Williamson [9] are concerned with prediction of sustained oscillation under zero forcing. The interesting work of Sandberg [8] involved nonlinear systems which contain Lipschitz continuous nonlinearities. Holtzman [6] treats a problem similar to Sandberg's using an entirely different method of analysis. The results presented here differ from all of the results mentioned above since we require nonzero forcing, and we allow nonlinear elements which may be discontinuous (e.g., relays with or without dead zones and/or hysteresis). Moreover our theorems differ in kind since we state that *if one can solve the approximate describing function problem and if certain bounds*, which depend on the exact form of the solution of the describing function problem, *can be satisfied, then there is a periodic solution of the exact problem.*

Other somewhat related work is contained in the paper of Popov [7]. A good elementary discussion of describing functions can be found in the book of Gelb and Van der Velde [4]. For background on oscillation theory see Hale [5].

2. AN ABSTRACT RESULT

We first study the periodic solutions of a nonlinear system of the form

$$x = h + GF(x) \tag{E}$$

in the space $X = \{z : R^1 \to R^1 : z \text{ continuous, T-periodic, } z(t + T/2) = -z(t)\}$ with norm $\| z \| = \sup\{|z(t)| : 0 \leq t \leq T\}$. If $w = 2\pi/T$, then

$$z = \underset{N\to\infty}{\text{l.i.m.}} \left(\Sigma_{\substack{|n|\leq N\\ n \text{ odd}}} z_n \exp(inwt)\right)$$

where

$$z_n = T^{-1} \int_0^T z(t)\exp(-inwt)dt$$

are the Fourier coefficients, $z_{-n} = \overline{z}_n$ and $n = \pm 1, \pm 3, \pm 5, \ldots$. We make the convention here and later that Σ' means sum over all odd integers. Thus for $z \in X$, Parseval's identity is

$$T^{-1} \int_0^T z(t)^2 \, dt = \Sigma' |z_n|^2 \tag{N}$$

Define a projection P on X by

$$(Pz)(t) = z_1 \exp(iwt) + z_{-1} \exp(-iwt)$$

Define $B(y,\varepsilon) = \{z \in X : \| z - y \| \leq \varepsilon\}$, i.e. the closed ball about z of radius ε. Let X_2 denote the completion of X under the L^2-norm (N).

System (E) will be studied by first analyzing the approximating system

$$\overline{x} = h + GPF(\overline{x}) \tag{$\overline{E}$}$$

under the following assumptions.

(A1) $h(t) = a \sin(wt + b)$ for some real constants a and b, $a \neq 0$.

(A2) $(\bar{E})$ has a solution $\bar{x} = c_1 \exp(iwt) + c_{-1}\exp(-iwt)$.

(A3) $F : X \to X$ is continuous.

(A4) G is a linear map on X to X defined by

$$(Gz)(t) = \Sigma'(G_n z_n)\exp(inwt)$$

where $G_{-n} = \bar{G}_n$ and $\sup\{|nG_n| : n \text{ odd}\} = \gamma < \infty$.

We remark that if (A4) is true, it can be shown that $G : X \to X$ is completely continuous, indeed even that G maps the Hilbert space X_2 into X completely continuously. Note also that G commutes with the projection P. The following result will be needed later.

LEMMA 1. If (A1)-(A4) are true and if there are constants ε and $M > 0$ such that for all $x \in B(\bar{x},\varepsilon)$ one has $\| F(x) \| \leq M$ and

$$\| GP(F(x) - F(\bar{x})) + (I - P)GF(x) \| \leq \varepsilon$$

then equation (E) has a solution $x \in B(\bar{x},\varepsilon)$.

Proof. Apply the Schauder fixed point theorem.

Now consider the case where F is not necessarily continuous. Assume $F : X \to X_2$. We will approximate F by a sequence of continuous operators $F_m : X \to X$ such that:

(A5) for any point $z \in PX$ and any sequence $\{z_m\} \subset PX$, if $z_m \to z$, then $PF_m(z_m) \to PF(z)$.

We look at the corresponding problems

$$x_m = h + GF_m(x_m) \qquad (E_m)$$

and

$$\bar{x}_m = h + GPF_m(\bar{x}_m) \qquad (\bar{E}_m)$$

We say that (E) has a *weak solution* $x \in X$ if there is a sequence of equations (E_m) satisfying (A5), solutions $x_m \in X$ and a function $w \in X_2$ such that $F_m(x_m) \to w$ weakly in X_2, $x_m \to x$ in X and $x = h + Gw$. We will see below that for n^{th} order differential equations containing a nonlinear element, our weak solutions become solutions in the sense of Filippov [3].

We can now state the main result of this section.

THEOREM 1. Suppose (E_m) and $(\overline{E}_m)$ satisfy (A1-A4) for each $m \geq 1$ and $\{F_m\}$ satisfies (A5). If $\sup_m \| PF_m(\overline{x}_m) \| < \infty$, then $(\overline{E})$ has a solution $\overline{x}$. If in addition there are positive constants M, ε and ε_1 such that $0 < \varepsilon_1 < \varepsilon$ and for all $m \geq 1$ and all $x \in B(\overline{x},\varepsilon)$ we have

$$\| F_m(x) \| \leq M, \qquad \| GP(F_m(x) - F_m(\overline{x})) + (I - P)GF_m(x) \| \leq \varepsilon_1$$

then (E) has a weak solution x with $\| x - \overline{x} \| \leq \varepsilon$.

Proof. Since the sequence $PF_m(\overline{x}_m)$ is bounded, we may assume that there is a $z \in X_2$ such that $PF_m(\overline{x}_m) \to z$ weakly. Since G is completely continuous, then $GPF_m(\overline{x}_m) \to Gz$ in X and $\overline{x}_m \to h + Gz$. By (A5) it follows that $PF_m(\overline{x}_m) \to PF(h + Gz)$. Thus $h + Gz = \overline{x}$ and $\overline{x}$ solves (E).

From (A5) it follows that $PF_m(\overline{x}) \to PF(\overline{x})$ and so there exists $\varepsilon_2 \in (\varepsilon_1,\varepsilon)$ such that

$$\| GP(F_m(x) - F(\overline{x}_m)) + (I - P)GF_m(x) \| \leq \varepsilon_2$$

for all $x \in B(\overline{x}_m,\varepsilon_2)$ and for all m sufficiently large. By Lemma 1 above it follows that (E_m) has a solution $x_m \in B(\overline{x}_m,\varepsilon_2) \subset B(\overline{x},\varepsilon)$ for m sufficiently large.

Since $F_m(x_m)$ is a bounded sequence in X, it is weakly precompact in X_2. By possibly taking a subsequence we can assume that $F_m(x_m) \to w$ (weakly) for some $w \in X_2$. Since G is completely continuous, then $GF_m(x_m) \to Gw$ in X. Thus

$$x_m = h + GF_m(x_m) \to h + Gw$$

If we define $x = h + Gw$, then the proof is complete.

The next result follows immediately from Theorem 1.

COROLLARY 1. Suppose (E_m) and $(\overline{E}_m)$ satisfy (A1-A4) for each $m \geq 1$ and $\{F_m\}$ satisfies (A5). If $\sup_m \| PF_m(\overline{x}_m) \| < \infty$, then $(\overline{E})$ has a solution $\overline{x}$. If in addition there exist positive constants M, ε and ε_1 such that for all $m \geq 1$ and all $x \in B(\overline{x},\varepsilon)$ we have $\| F_m(x) \| \leq M$ and

$$\| GP(F_m(x) - F_m(\overline{x})) \| + M\Big(\sum_{\substack{|n|\geq 3 \\ n \text{ odd}}} |G_n|^2\Big)^{1/2} \leq \varepsilon_1 < \varepsilon$$

then (E) has a weak solution $x \in B(\overline{x},\varepsilon)$.

3. DIFFERENTIAL EQUATIONS

Let $Ly = \Sigma_{j=0}^{J} \alpha_j y^{(j)}$ be a linear differential operator with constant coefficients, $\alpha_J \neq 0$ and $J \geq 1$. Define $p(s) = \Sigma_{j=0}^{J} \alpha_j s^j$ so that $p(s)^{-1}$ is the transfer function associated with L. Consider the nonlinear differential equation

$$Ly + f(y + a \sin(wt + b)) = 0$$
$$y(0) = \xi_0, y'(0) = \xi_1, \ldots, y^{(J-1)}(0) = \xi_{J-1} \tag{1}$$

under the assumptions:

(A6) $f(y)$ is an odd, piecewise C^1 function on R^1 to R^1 with a finite set of jump discontinuities at the points $\{x_j\}$. Moreover there exist non-negative constants f_0 and f_1 such that $|f(y)| \leq f_0 + f_1|y|$ for all $y \geq 0$.

(A7) $p(inw) \neq 0$ for $n = \pm 1, \pm 3, \pm 5, \ldots$.

Define $h(t) = a \sin(wt + b)$, $x = y + h$ and $F(z)(t) \equiv -f(z(t))$ for all $z \in X$. Then (1) becomes

$$L(x - h) = F(x) \tag{1'}$$

Clearly (1') has a T-periodic solution (with $w = 2\pi/T$) if and only if (E) has a solution in X when G is the map defined by (A4) with $G_n = p(inw)^{-1}$ for all n.

The projection P of section II determines the usual describing function approximation $N(A)$ to the nonlinearity f, i.e.,

$$PF(A \sin wt) = N(A)A \exp(iwt) + \overline{N(A)}A \exp(-iwt)$$

In order to insure that $(\bar{E})$ will have a solution we assume:

(A8) There is an $A_0 > 0$ such that the curves $y = A/a$ and $y = |p(iw)|\ |p(iw) - N(A)|^{-1}$ are defined near A_0 and cross at A_0. This insures that $(\bar{E})$ will have a solution of the form $\bar{x} = A_0 \sin(wt + B_0)$ for some B_0. Also $|A_0 - x_j| > \varepsilon$ for all j.

We define a sequence $f_m(x)$ as follows. We can assume that the points x_j of discontinuity of f are numbered so that $x_1 < x_2 < x_3 < \ldots < x_k$. For any $m \geq 1$ and any point x_j define $\delta_{jm} = \text{minimum}\{(x_{j+1} - x_j)/4, (x_j - x_{j-1})/4, 1/m\}$ (where $x_0 = x_1 - 1$ and $x_{k+1} = x_k + 1$). Define

$$f_m(x) = \begin{cases} f(x) & \text{if} \quad |x - x_j| \geq \delta_{jm} \quad \text{for} \quad 1 \leq j \leq k \\ \text{linear in the regions } |x - x_j| \leq \delta_{jm} \end{cases}$$

Define $F_m(z)(t) \equiv -f_m(z(t))$. Let $N_m(A)$ be the describing function for F_m. Then $N_m(A) \to N(A)$ as $m \to \infty$ uniformly for A on compact subsets of the interval $(0,\infty)$. Therefore (A8) is also true for $N_m(A)$ when m is sufficiently large and $(\overline{E}_m)$ will have solutions $\overline{x}_m = A_m \sin(wt + B_m) \to \overline{x}$ as $m \to \infty$. Thus the next result follows immediately from Corollary 1 above.

THEOREM 2. If (A6-A8) are true, if $M(\varepsilon) = f_0 + f_1(\varepsilon + A_0)$ and if there exist positive constants ε_1 and ε such that for all m sufficiently large and for all $x \in B(\overline{x},\varepsilon)$

$$\| GP(F_m(x) - F_m(\overline{x})) \| + M(\varepsilon)\Big(\sum_{\substack{|n|\geq 3 \\ n \text{ odd}}} |G_n|^2\Big)^{1/2} \leq \varepsilon_1 < \varepsilon \tag{2}$$

then there exists $x \in X$ with $\| x - \overline{x} \| \leq \varepsilon$ such that x is a solution of (1') in the sense of Filippov.

We note that hypotheses (A6) could be generalized to include nonlinear terms with hysteresis. Theorem 2 could also be modified to allow nonlinear terms which do not satisfy the bound $|f(y)| \leq f_0 + f_1|y|$. However a very large class of useful nonlinear elements do satisfy this inequality, often with $f_1 = 0$. Assumption (A7) is the usual requirement that the linear part of the system does not admit as a solution any subharmonics. As we have already pointed out, (A8) insures that the approximating system $(\overline{E})$ determined when f is replaced by its describing function should have a solution. The remaining hypothesis (2) divides into two parts. The first term in (2) is small when F is not "too nonlinear." The second term is small when the linear operator L determines a good low pass filter. All of our hypotheses are normal, natural and expected in typical engineering applications.

As an example we consider the nonlinear, discontinuous relay with hysteresis shown in Figure 1. The continuous approximations f_m are shown in Figure 2. The describing function for f has the form

$$N(A) = \frac{4D}{\pi A}\{(1 - (\frac{\delta}{A})^2)^{1/2} - i(\frac{\delta}{A})\}, \qquad A \geq \delta$$

Note that f(z) is not defined when $\| z \| < \delta$ and N(A) is not defined when $0 < A < \delta$.

In order to find a solution of $(\overline{E})$ we assume that $|1 + G_1N(\delta)| < |a|/\delta$. This insures that

$$|1 + G_1N(A)| = |a|/A \tag{3}$$

has a solution $A_0 > \delta$. A worse case analysis shows that $\| P(F(x) - F(\overline{x})) \|$ can be computed as follows. The cosine coefficient is $(4D\varepsilon)/(\pi A)$ while the sine coefficient is the larger of

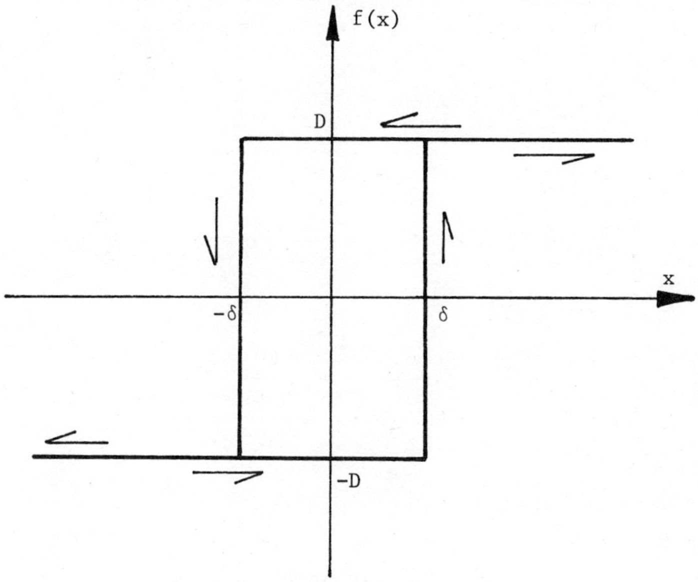

Figure 1. Relay with Hysteresis

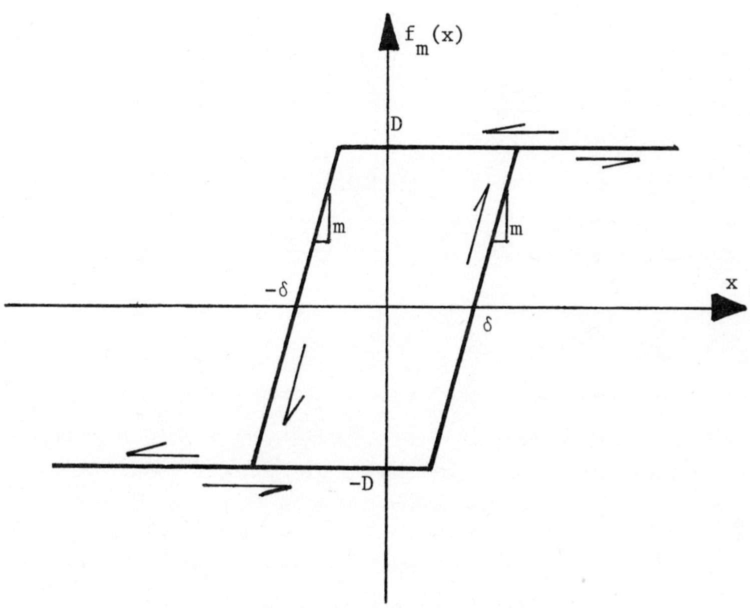

Figure 2. Continuous Approximation of Relay with Hysteresis

$$\frac{4D}{\pi}\int_{\sin^{-1}(\frac{\delta-\varepsilon}{A})}^{\sin^{-1}(\frac{\delta}{A})} \sin t \, dt \qquad \text{or} \qquad \frac{4D}{\pi}\int_{\sin^{-1}(\frac{\delta}{A})}^{\sin^{-1}(\frac{\delta+\varepsilon}{A})} \sin t \, dt$$

for $0 < \varepsilon < \delta$. Thus for m large, $\| P(F_m(x) - F_m(\bar{x})) \|$ is as close as we like to

$$E(\delta,D,\varepsilon,A) = \frac{4D}{\pi A}\left[\varepsilon^2 + \left\{\frac{2\delta\varepsilon + \varepsilon^2}{(A^2 - (\delta + \varepsilon)^2)^{1/2} + (A^2 - \delta^2)^{1/2}}\right\}^2\right]^{1/2}$$

Condition (2) will be true when

$$|G_1|E(\delta,D,\varepsilon,A) + D\Big(\sum_{\substack{n \text{ odd} \\ |n|\geq 3}} |G_n|^2\Big)^{1/2} < \varepsilon \tag{4}$$

As a specific illustration take $D = a = w = 1$, $b = 0$, $\delta = 0.1$ and $p(s) = s(s + 1)^2$. The differential equation is

$$y''' + 2y'' + y' + f(y + \sin t) = 0 \tag{5}$$

Condition (3) is equivalent to

$$\phi(A) = (A - (2/\pi)(1 - .01A^{-2})^{1/2})^2 + .04(\pi^2 A^2)^{-1} = 1$$

This equation has one and only one root $A_0 \cong 1.6347$. Since

$$\Big(\sum_{\substack{n \text{ odd} \\ |n|\geq 3}} |G_n|^2\Big)^{1/2} \cong .048605$$

then (4) reduces to

$$.5\ E(.1,1,\varepsilon,A_0) + .048605 < \varepsilon$$

A straightforward computation will verify that $\varepsilon = .07979$ will do. Thus the amplitude A_0 predicted by the describing function method and the actual amplitude of the periodic solution of (5) differ by an amount with relative error at most $\varepsilon/A_0 = 4.88\%$.

ACKNOWLEDGMENT

Both authors were supported by the National Science Foundation under Grant ENG-77-28446. Professor Michel was additionally supported by the Engineering Research Institute, Iowa State University, Ames, Iowa.

REFERENCES

1. R. W. Bass, Mathematical Legitimacy of Equivalent Linearization by Describing Functions, *Automatic and Remote Control*, ed. F. J. Coales, London, Butterworths, 1961, 895-905.

2. A. R. Bergen and R. L. Franks, Justification of the Describing Function Method, *SIAM J. Control*, 9(1971), 586-589.

3. A. F. Filippov, Differential Equations with Discontinuous Right Hand Sides (in Russian), *Math. Sb.* 5(1960), 99-128 (English Transl., *Amer. Math. Soc. Transl.* 42(1964), 199-231).

4. A. Gelb and W. E. Vander Velde, *Multiple-Input Describing Functions and Nonlinear System Design*, McGraw-Hill Book Co., New York, 1968.

5. J. K. Hale, *Oscillations in Nonlinear Systems*, McGraw-Hill, New York, 1963.

6. J. M. Holtzman, Contraction Maps and Equivalent Linearization, *Bell System Tech. J.*, 46(1967), 2405-2435.

7. V. M. Popov, Absolute Stability of Nonlinear Systems of Automatic Control, *Automation and Remote Control*, 22(1961), 857-875.

8. I. W. Sandberg, On the Response of Nonlinear Control Systems to Periodic Input Signals, *Bell System Tech. J.*, 43(1964), 911-926.

9. Darrell Williamson, Describing Function Analysis and Oscillation in Nonlinear Networks, *Int. J. Control*, 24(1976), 283-296.

A DETERMINISTIC DIFFUSIVE EPIDEMIC MODEL WITH AN INCUBATION PERIOD

G. F. Webb

Mathematics Department
Vanderbilt University
Nashville, Tennessee

1. INTRODUCTION

This work continues the study of deterministic diffusive epidemic models begun in [13]. In the present study our purpose is to modify the model investigated in [13] to allow for a period of incubation of the infection in exposed individuals. In the model we study here infected individuals remain in an exposed class for a fixed period of time during which they acquire a sufficient level of infectiousness to become infective to susceptible members of the population. These individuals then remain infective for a fixed period of time after which they acquire permanent immunity. Models incorporating incubation periods are typically associated with measles and sometimes called measles models. Such models have been studied by a number of researchers (including H. Soper [11], L. Wilson [15], P. Waltman [12], and F. Hoppensteadt [4]) in the case of populations that are spatially homogeneous. The contribution we make here to the investigation of such models is to allow the spatial inhomogeneity and diffusion of the epidemic population and its subclasses. Accordingly, we are led to a mathematical model involving nonlinear partial differential equations. In fact, since our epidemic model allows fixed time delays before individuals move from one subclass to another, our model of an epidemic involves nonlinear functional partial differential equations or nonlinear partial differential equations with delay.

We suppose that the population is divided into four classes: the susceptible class (S) (individuals not infected but capable of becoming infected), the exposed class (E) (individuals infected but not yet infective), the infective class (I) (individuals capable of transmitting the disease to members of the susceptible class, and the removed class (R) (individuals who have died or who have been quarantined or who have recovered with permanent immunity). We suppose that members of the susceptible class enter the exposed class at a rate proportional to the product of the sizes of the classes (S) and (I) (with constant of proportionality $r > 0$). We suppose that individuals remain in the exposed class for a fixed period of time $\tau > 0$ and that individuals who survive as infectives remain in the infective class for a fixed period of time $\sigma > 0$. We suppose also that individuals who do not survive as infectives (that is, who have either died or been quarantined) are removed from the infective class at a rate proportional to the size of the class (I) (with constant of proportionality $\lambda \geq 0$). Finally, we suppose that diffusion of each of the classes (S), (E), (I), and (R) occurs within the population and that for each class this diffusion is approximated by the product of a positive coefficient of diffusion and the Laplacian of the density function for that class. For the sake of simplicity we require that the spatial region in which the epidemic occurs is a one-dimensional interval $[0,L]$ and that the population is constrained to remain within this interval for all time. We note, however, that our methods carry over to the case of spatial regions in higher dimensions.

Let $S(x,t)$, $E(x,t)$, $I(x,t)$, and $R(x,t)$ denote the population density functions of the classes (S), (E), (I), and (R), respectively, where x is a point in $[0,L]$ and t is the time evolved. Our model then has the form

$$S_t(x,t) = aS_{xx}(x,t) - rS(x,t)I(x,t) \qquad 0 \leq x \leq L, \qquad t \geq 0 \tag{1.1}$$

$$\begin{aligned} E_t(x,t) &= bE_{xx}(x,t) + rS(x,t)I(x,t) - rS(x,t - \tau)I(x, t - \tau) \\ &\quad 0 \leq x \leq L, \qquad t \geq 0 \end{aligned} \tag{1.2}$$

$$\begin{aligned} I_t(x,t) &= cI_{xx}(x,t) + rS(x,t - \tau)I(x,t - \tau) \\ &\quad -rS(x,t - \tau - \sigma)I(x,t - \tau - \sigma) - \lambda I(x,t) \qquad 0 \leq x \leq L, \\ &\quad t \geq 0 \end{aligned} \tag{1.3}$$

$$R_t(x,t) = dR_{xx}(x,t) + rS(x,t-\tau-\sigma)I(x,t-\tau-\sigma) + \lambda I(x,t) \quad 0 \le x \le L,\ t \ge 0 \tag{1.4}$$

$$S_x(0,t) = E_x(0,t) = I_x(0,t) = R_x(0,t) = S_x(L,t) = E_x(L,t) = I_x(L,t) = R_x(L,t) = 0, \quad t \ge 0 \tag{1.5}$$

$$S(x,t) = \phi(x,t), \quad E(x,t) = \chi(x,t), \quad I(x,t) = \psi(x,t), \quad R(x,t) = \Gamma(x,t), \quad 0 \le x \le L, \quad -\tau-\sigma \le t \le 0 \quad \text{(where } \phi, \chi, \psi, \text{ and } \Gamma \text{ are prescribed initial values on } [0,L]\times[-\tau-\sigma,0]) \tag{1.6}$$

The coefficients of diffusion a, b, c, and d are all taken as positive. We observe that once $S(x,t)$ and $I(x,t)$ are known, $E(x,t)$ and $R(x,t)$ can be determined.

Our goal, therefore, is to solve equations (1.1) and (1.3) subject to the boundary and initial conditions in (1.5) and (1.6). We shall prove the existence of solutions to this problem and analyze their large time behavior. We may summarize our results in the following way: (i) if ϕ, ψ are continuous on $[0,L] \times [-\tau-\sigma,0]$, then mild solutions of (1.1) and (1.3) exist for all time; (ii) if ϕ, ψ satisfy additional smoothness conditions, then these mild solutions are strong solutions (in fact, classical solutions); (iii) if ϕ, ψ satisfy certain positivity assumptions, then the solutions of (1.1) and (1.3) are positive for all time; and (iv) if the solutions $S(t)$, $I(t)$ of (1.1) and (1.3) remain positive for all time, then the limits $\lim_{t\to\infty} S(x,t) \overset{\text{def}}{=} S_\infty(x)$ and $\lim_{t\to\infty} I(x,t) \overset{\text{def}}{=} I_\infty(x)$ exist uniformly in x in $[0,L]$, S_∞ and I_∞ are constant functions of x on $[0,L]$, and further, if $\lambda = 0$ then $I_\infty = L^{-1}(\int_0^L \psi(x,0)dx - r\int_{-\tau-\sigma}^{-\tau} \int_0^L \phi(x,s)\psi(x,s)\,dxds)$, $S_\infty = 0$, and if $\lambda > 0$ then $I_\infty = 0$, $S_\infty > 0$.

The methods we use to prove the facts in (i)-(iv) above draw heavily upon the theory of semigroups of linear operators in Banach spaces. Accordingly, we suppose upon the reader an acquaintance with this theory as presented in [7] or [9]. The methods we use here are somewhat different (and simpler and more direct) then the methods we used in the earlier study [13]. One difference is the method we use to prove the existence of the limits $\lim_{t\to\infty} S(x,t) = S_\infty$, $\lim_{t\to\infty} I(x,t) = I_\infty$, which relies on a device of C. Kahane in [6]. In Section 2 we establish some results for the spatially homogeneous case which are, in fact, analogous to the results in (i)-(iv) for the spatially inhomogeneous case. Although these results for the spatially homogeneous case are known, we present them here because it is instructive to do so before analyzing the spatially inhomogeneous case. In Section 3 we prove the facts in (i), (ii), and (iii), and in Section 4 we prove the facts in (iv).

2. THE SPATIALLY HOMOGENEOUS CASE

The equations for the epidemic model in this case have the form

$$S'(t) = -rS(t)I(t), \qquad t \geq 0 \tag{2.1}$$

$$I'(t) = rS(t-\tau)I(t-\tau) - rS(t-\tau-\sigma)I(t-\tau-\sigma) - \lambda I(t), \quad t \geq 0 \tag{2.2}$$

$$S(t) = \phi(t), \qquad I(t) = \psi(t), \qquad -\tau-\sigma \leq t \leq 0 \tag{2.3}$$

We state and prove below some known results for the problem (2.1)-(2.3). The reader is referred to [12, Chapter 2] and [4, Chapter III] for extensive treatments of this model and similar models.

PROPOSITION 2.1. Let r, τ, σ be positive constants, let λ be a nonnegative constant, let φ, ψ be continuous functions from $[-\tau-\sigma,0]$ to $[0,\infty)$, and let

$$\psi(0) - r\int_{-\tau-\sigma}^{-\tau}\phi(\theta)\psi(\theta)d\theta > 0 \tag{2.4}$$

Then there exist unique functions S, I from $[-\tau-\sigma,\infty)$ to $[0,\infty)$ satisfying (2.1)-(2.3). Further, the limits $\lim_{t\to\infty} S(t) \overset{def}{=} S_\infty$ and $\lim_{t\to\infty} I(t) \overset{def}{=} I_\infty$ exist and

$$\text{if } \lambda = 0, \quad \text{then } S_\infty = 0 \quad \text{and} \quad I_\infty = \psi(0) - r\int_{-\tau-\sigma}^{-\tau}\phi(\theta)\psi(\theta)d\theta \tag{2.5}$$

$$\text{if } \lambda > 0, \quad \text{then } S_\infty = \phi(0)\exp[-r\lambda^{-1}(\psi(0) - r\int_{-\tau-\sigma}^{-\tau}\phi(\theta)\psi(\theta)d\theta)] \text{ and } I_\infty = 0 \tag{2.6}$$

Proof. From (2.1)-(2.3) we obtain

$$S(t) = \phi(0)\exp[-r\int_0^t I(s)ds], \qquad t \geq 0 \tag{2.7}$$

$$I(t) = \psi(0) - r\int_{-\tau-\sigma}^{-\tau}\phi(\theta)\psi(\theta)d\theta + r\int_{t-\tau-\sigma}^{t-\tau} S(s)I(s)ds - \lambda\int_0^t I(s)ds, \qquad t \geq 0 \tag{2.8}$$

The method of steps may be used to solve (2.7) and (2.8) simultaneously. That is, solve (2.8) for I(t), $0 \leq t \leq \tau$, and then S(t) is defined by (2.7) for $0 \leq t \leq \tau$. Next, solve (2.8) for I(t), $\tau \leq t \leq 2\tau$, and then S(t) is defined by (2.7) for $\tau \leq t \leq 2\tau$. Continue this process to obtain S(t),

$I(t)$ for all $t \geq 0$. Notice that (2.7) implies that $S(t) \geq 0$ for all $t \geq 0$ and then (2.4) and (2.8) imply that $I(t) \geq 0$ for all $t \geq 0$.

To establish (2.5) and (2.6) we see first from (2.7) that $S(t)$ is non-increasing, and hence $\lim_{t\to\infty} S(t) = S_\infty$ exists. Integrate (2.1) and add the result to (2.8) to obtain

$$\begin{aligned} S(t) + I(t) &+ r\int_{-\tau-\sigma}^{-\tau}\phi(\theta)\psi(\theta)d\theta + r\int_0^{t-\tau-\sigma}S(s)I(s)ds \\ &+ r\int_{t-\tau}^{t}S(s)I(s)ds + \lambda\int_0^t I(s)ds = \phi(0) + \psi(0), \qquad t \geq 0 \end{aligned} \tag{2.9}$$

Then (2.9) implies that $\int_0^\infty S(s)I(s)ds < \infty$ and $\lim_{t\to\infty} \int_{t-\tau}^{t}S(s)I(s)ds = 0$. If $\lambda = 0$, then (2.8) yields that $\lim_{t\to\infty} I(t) = I_\infty$ exists and $I_\infty = \psi(0) - r\int_{-\tau-\sigma}^{-\tau}\phi(\theta)\psi(\theta)d\theta$. Thus, $\int_0^\infty I(s)ds = \infty$ by (2.4), and so by (2.7) $\lim_{t\to\infty} S(t) = S_\infty = 0$. If $\lambda > 0$, then (2.9) implies that $\int_0^\infty I(s)ds < \infty$. Since $\lim_{t\to\infty} \int_{t-\tau-\sigma}^{t-\tau} S(s)I(s)ds = 0$, (2.8) yields that $\lim_{t\to\infty} I(t) = I_\infty$ exists. Since $\int_0^\infty I(s)ds < \infty$, we must have $I_\infty = 0$. Then (2.8) yields that $\lambda\int_0^\infty I(s)ds = \psi(0) - r\int_{-\tau-\sigma}^{-\tau}\phi(\theta)\psi(\theta)d\theta$. From (2.7) we thus obtain the formula for S_∞ given in (2.6).

3. EXISTENCE OF SOLUTIONS IN THE SPATIALLY INHOMOGENEOUS CASE

Let X denote the Banach space $C([0,L]; R)$ of continuous real-valued functions ϕ on $[0,L]$ with supremum norm $\|\phi\| = \sup_{0\leq x\leq L} |\phi(x)|$. Let $V(\phi) = \int_0^L \phi(x)dx$ for all ϕ in X. Let X^+ denote the subset of X consisting of nonnegative-valued functions in X. Let A denote the linear operator from $D(A) \subset X$ to X defined by

$$\begin{aligned} &A\phi = d^2/dx^2\, \phi \\ &D(A) = \{\phi \in X\colon\ \phi \text{ is twice continuously differentiable on } \\ &\qquad [0,L] \text{ and } d/dx\,\phi(0) = d/dx\,\phi(L) = 0\} \end{aligned} \tag{3.1}$$

The following facts concerning the operator A are well-known:

$$\begin{aligned} &A \text{ is the infinitesimal generator of a strongly continuous} \\ &\quad \text{analytic semigroup } e^{tA},\ t \geq 0 \text{ of linear operators in} \\ &\quad X \text{ satisfying } \|e^{tA}\phi\| \leq \|\phi\| \text{ for all } \phi \text{ in } X \text{ and} \\ &\quad t \geq 0 \end{aligned} \tag{3.2}$$

$$\begin{aligned} e^{tA}\phi(x) &= 2L^{-1}\int_0^L G(x,\xi,t)\phi(\xi)d\xi + L^{-1}\int_0^L \phi(\xi)d\xi, \text{ where } G(x,\xi,t) \\ &= \Sigma_{n=1}^{\infty} \cos n\pi\xi L^{-1}\cos n\pi x L^{-1}\exp[-(n\pi L^{-1})^2 t] \text{ and } |G(x,\xi,t)| \leq \\ &\quad L(4\pi t)^{-1/2} \text{ for all } x \text{ in } [0,L],\ \xi \text{ in } [0,L], \qquad t > 0 \end{aligned} \tag{3.3}$$

$$V(e^{tA}\phi) = V(\phi) \qquad \text{for all } \phi \text{ in } X, \qquad t \geq 0 \tag{3.4}$$

$$e^{tA}(X^+) \subset X^+ \qquad \text{for all} \qquad t \geq 0 \tag{3.5}$$

It is not possible to obtain solutions of equations (1.1) and (1.3) for arbitrary initial functions ϕ, ψ in X, and therefore we shall prove the existence of mild solutions, that is, solutions to the integrated version of these equations:

$$S(t) = e^{taA}\phi(0) - r\int_0^t e^{(t-s)aA}S(s)I(s)ds, \qquad t \geq 0 \tag{3.6}$$

$$\begin{aligned} I(t) &= e^{tcA}\psi(0) + r\int_0^t e^{(t-s)cA}[S(s-\tau)I(s-\tau) - \\ &\quad S(s-\tau-\sigma)I(s-\tau-\sigma)]ds - \lambda\int_0^t e^{(t-s)cA}I(s)ds, \qquad t \geq 0 \\ &= e^{t(cA-\lambda)}\psi(0) + r\int_0^t e^{(t-s)(cA-\lambda)}[S(s-\tau)I(s-\tau) \\ &\quad - S(s-\tau-\sigma)I(s-\tau-\sigma)]ds, \qquad t \geq 0 \end{aligned} \tag{3.7}$$

$$S(t) = \phi(t), \qquad I(t) = \psi(t), \qquad -\tau-\sigma \leq t \leq 0 \tag{3.8}$$

PROPOSITION 3.1. Let r, τ, σ, a, c be positive constants, let λ be a nonnegative constant, and let ϕ, ψ be continuous functions from $[-\tau-\sigma, 0]$ to X. There exists a unique pair of continuous functions S, I from $[0,\infty)$ to X satisfying (3.6), (3.7), and (3.8).

Proof. We will use the mtheod of steps to establish the existence and uniqueness of solutions to this problem. The nonhomogeneous linear equation (3.7) has a unique continuous solution I(t) for $0 \leq t \leq \tau$ by [7, Theorem 1.19, p. 486]. Now substitute I(t) into equation (3.6) for $0 \leq t \leq \tau$ and equation (3.6) takes the form

$$S(t) = e^{taA}\phi(0) + \int_0^t e^{(t-s)aA}B(s,S(s))ds \tag{3.9}$$

where the mapping B from $[0,\tau] \times X$ to X is given by $B(s,x) = -rxI(s)$. Since B(s,x) is continuous in s and x jointly and linear in x, (3.9) has a unique continuous solution S(t) for $0 \leq t \leq \tau$ by Theorem 2.1, p. 235 in [9]. Now S(t), I(t) are defined for $0 \leq t \leq \tau$ and we may repeat this process to obtain unique continuous solutions S(t), I(t) to (3.6)-(3.7) on $[\tau, 2\tau]$, $[2\tau, 3\tau], \ldots,$ and hence on $[0,\infty)$.

Since the solutions we seek will have physical meaning only if they have nonnegative values, we give below the following sufficient conditions on ϕ and ψ to guarantee the nonnegativity of S(t) and I(t):

PROPOSITION 3.2. Let r, τ, σ, a, c be positive constants, let λ be a non-negative constant, let ϕ, ψ be continuous functions from $[-\tau-\sigma,0]$ to X^+, and let ϕ, ψ satisfy

$$e^{t(cA - \lambda)}\psi(0) \geq r\int_{-\tau-\sigma}^{0} e^{(t - s - \tau - \sigma)(cA - \lambda)}\phi(s)\psi(s)ds, \quad t \geq 0 \tag{3.10}$$

$$\begin{aligned} e^{t(cA - \lambda)}\psi(0) &+ r\int_{-\tau}^{0} e^{(t - s - \tau)(cA - \lambda)}\phi(s)\psi(s)ds \\ &\geq r\int_{-\tau-\sigma}^{0} e^{(t - s - \tau - \sigma)(cA - \lambda)}\phi(s)\psi(s)ds \\ &+ r\{\sqrt{2c^{-1}}\,[V(\phi(0)) + 2\,\|\,\phi(0)\,\|V(\phi(0) + \psi(0))] \\ &+ V(\phi(0))L^{-1}\}, \qquad t \geq 0 \end{aligned} \tag{3.11}$$

The solutions S(t), I(t) of (3.6)-(3.8) satisfy S(t), I(t) $\in X^+$ for all $t \geq 0$.

Proof. First, we show that the solution S(t) of (3.6) lies in X^+ for all $t \geq 0$. To establish this fact we will apply Proposition 4.2, p. 353 in [9]. We need only to show that for $(t,x) \in [0,\infty) \times X^+$ fixed,

$$\lim_{h\to 0^+} \inf_{z\in X^+} \|\, x + hB(t,x) - z \,\| = o(h)$$

where B(t,x) is as in (3.9). But if $(t,x) \in [0,\infty) \times X^+$, then $x + hB(t,x) = (1 - hrI(t))x \in X^+$ for h sufficiently small, and so the result follows immediately.

Next, we will use the facts that $\phi(t), \psi(t) \in X^+$ for $-\tau-\sigma \leq t \leq 0$, $S(t) \in X^+$ for $t \geq 0$, (3.10), and (3.11) to show that $I(t) \in X^+$ for all $t \geq 0$. We first establish some additional facts that we will need. First,

$$\text{if } t > 0 \text{ and } I(s) \in X^+ \text{ for } 0 \leq s \leq t, \quad \text{then } \|\, S(t) \,\| \leq \|\, \phi(0) \,\| \tag{3.12}$$

The claim (3.12) follows immediately from (3.5) and (3.6). Next, we show that

$$\text{if } t > 0 \text{ and } I(s) \in X^+ \text{ for } 0 \leq s \leq t, \quad \text{then } \int_0^t V(S(s)I(s))ds \leq V(\phi(0)) \tag{3.13}$$

Since $S(s), I(s) \in X^+$ for $0 \leq s \leq t$, (3.13) follows immediately from (3.4), (3.6), and the fact that

$$V(S(t)) = V(\phi(0)) - r\int_0^t V(S(s)I(s))ds$$

Next, we show that

$$\text{if } t > 0 \text{ and } I(s) \in X^+ \text{ for } 0 \leq s \leq t, \quad \text{then } V(I(t)) \leq V(\phi(0) + \psi(0)) \tag{3.14}$$

To establish (3.14) use (3.4), (3.6), (3.7), and the fact that

$$V(S(t)) + V(I(t)) = V(\phi(0)) + V(\psi(0)) - r\int_{t-\tau}^{t} V(S(s)I(s))ds$$
$$- r\int_{0}^{t-\tau-\sigma} V(S(s)I(s))ds - r\int_{-\tau-\sigma}^{-\tau} V(S(s)I(s))ds$$
$$- \lambda\int_{0}^{t} V(I(s))ds$$

Next, we show that

$$\text{if } t > 0 \text{ and } I(s) \in X^+ \text{ for } 0 \leq s \leq t, \text{ then}$$
$$\left\| \int_0^t e^{(t-s)cA} S(s)I(s)ds \right\| \leq \sqrt{2c^{-1}}[V(\phi(0)) \tag{3.15}$$
$$+ 2 \| \phi(0) \| V(\phi(0) + \psi(0))] + V(\phi(0))L^{-1}$$

To establish (3.15) we use (3.3), (3.12), (3.13), and (3.14) to argue that

$$\left\| \int_0^t e^{(t-s)cA} S(s)I(s)ds \right\|$$
$$= \sup_{0\leq x\leq L} \left| \int_0^t e^{(t-s)cA} S(s)I(s)(x)ds \right|$$
$$= \sup_{0\leq x\leq L} \left| \int_0^t [2L^{-1}\int_0^L G(x,\xi,c(t-s))S(s)I(s)(\xi)d\xi \right.$$
$$\left. + L^{-1}\int_0^L S(s)I(s)(\xi)d\xi]ds \right|$$
$$\leq 2L^{-1}\int_0^t L(2c(t-s))^{-1/2} V(S(s)I(s))ds + L^{-1}\int_0^t V(S(s)I(s))ds$$
$$\leq \sqrt{2c^{-1}} [\int_0^{t-1} (t-s)^{-1/2} V(S(s)I(s))ds$$
$$+ \int_{t-1}^{t} (t-s)^{-1/2} V(S(s)I(s))ds] + L^{-1} V(\phi(0))$$
$$\leq \sqrt{2c^{-1}} [\int_0^{t-1} V(S(s)I(s))ds$$
$$+ \sup_{t-1\leq s\leq t} V(S(s)I(s))\int_{t-1}^{t} (t-s)^{-1/2}ds] + L^{-1} V(\phi(0))$$
$$\leq \sqrt{2c^{-1}} [V(\phi(0)) + \| \phi(0) \| V(\phi(0) + \psi(0))2] + L^{-1} V(\phi(0))$$

Now let n be a nonnegative integer such that $n\tau < \sigma \leq (n+1)\tau$. From (3.7) we have that for $0 \leq t \leq \tau$,

$$I(t) = e^{t(cA-\lambda)}\psi(0) + r\int_{-\tau}^{t-\tau} e^{(t-s-\tau)(cA-\lambda)}\phi(s)\psi(s)ds$$
$$- r\int_{-\tau-\sigma}^{t-\tau-\sigma} e^{(t-s-\tau-\sigma)(cA-\lambda)}\phi(s)\psi(s)ds$$

$$\geq e^{t(cA - \lambda)}\psi(0) - r\int_{-\tau-\sigma}^{-\sigma} e^{(t - s - \tau - \sigma)(cA - \lambda)}\phi(s)\psi(s)ds$$

(since $t - \tau \leq 0$ and $\phi(s)$, $\psi(s) \in X^+$ for $-\tau-\sigma \leq s \leq 0$)

≥ 0 (by (3.10))

For $\tau < t \leq 2\tau$ we have that

$$I(t) = e^{t(cA - \lambda)}\psi(0) + r\int_{-\tau}^{t-\tau} e^{(t - s - \tau)(cA - \lambda)}S(s)I(s)ds$$

$$- r\int_{-\tau-\sigma}^{t-\tau-\sigma} e^{(t - s - \tau - \sigma)(cA - \lambda)}S(s)I(s)ds$$

$$\geq e^{t(cA - \lambda)}\psi(0) - r\int_{-\tau-\sigma}^{t-\tau-\sigma} e^{(t - s - \tau - \sigma)(cA - \lambda)}S(s)I(s)ds$$

(since $t - \tau \leq \tau$ and $S(s)$, $I(s) \geq 0$ for $-\tau-\sigma \leq s \leq \tau$)

≥ 0 by (3.10) (if $n \geq 1$)

For $n\tau < t \leq (n + 1)\tau$ we have by induction that

$$I(t) = e^{t(cA - \lambda)}\psi(0) + r\int_{-\tau}^{t-\tau} e^{(t - s - \tau)(cA - \lambda)}S(s)I(s)ds$$

$$- r\int_{-\tau-\sigma}^{t-\tau-\sigma} e^{(t - s - \tau - \sigma)(cA - \lambda)}S(s)I(s)ds$$

$$\geq e^{t(cA - \lambda)}\psi(0) - r\int_{-\tau-\sigma}^{n\tau-\sigma} e^{(t - s - \tau - \sigma)(cA - \lambda)}S(s)I(s)ds$$

(since $t - \tau \leq n\tau$, $n\tau - \sigma < 0$ and by the induction hypothesis $S(s)$, $I(s) \geq 0$ for $-\tau-\sigma \leq s \leq n\tau$)

≥ 0 (by (3.10))

For $(n + 1)\tau < t \leq \tau + \sigma$ we have that

$$I(t) = e^{t(cA - \lambda)}\psi(0) + r\int_{-\tau}^{t-\tau} e^{(t - s - \tau)(cA - \lambda)}S(s)I(s)ds$$

$$- r\int_{-\tau-\sigma}^{t-\tau-\sigma} e^{(t - s - \tau - \sigma)(cA - \lambda)}S(s)I(s)ds$$

$$\geq e^{t(cA - \lambda)}\psi(0) - r\int_{-\tau-\sigma}^{0} e^{(t - s - \tau - \sigma)(cA - \lambda)}S(s)I(s)ds$$

(since $t - \tau \leq \sigma \leq (n + 1)\tau$ and $S(s)$, $I(s) \geq 0$ for $-\tau-\sigma \leq s \leq (n + 1)\tau$) ≥ 0 (by (3.10))

For $\tau + \sigma < t \leq 2\tau + \sigma$ we have that

$$I(t) = e^{t(cA - \lambda)}\psi(0) + r\int_{-\tau}^{0} e^{(t - s - \tau)(cA - \lambda)}\phi(s)\psi(s)ds$$

$$+ r\int_{0}^{t-\tau} e^{(t - s - \tau)(cA - \lambda)}S(s)I(s)ds$$

$$- r\int_{-\tau-\sigma}^{0} e^{(t - s - \tau - \sigma)(cA - \lambda)}\phi(s)\psi(s)ds$$

$$- r\int_{0}^{t-\tau-\sigma} e^{(t - s - \tau - \sigma)(cA - \lambda)}S(s)I(s)ds$$

$$\geq e^{t(cA - \lambda)}\psi(0) + r\int_{-\tau}^{0} e^{(t - s - \tau)(cA - \lambda)}\phi(s)\psi(s)ds$$

$$+ r\int_0^{t-\tau} e^{(t - s - \tau)(cA - \lambda)} S(s)I(s)ds$$

$$- r\int_{-\tau-\sigma}^{0} e^{(t - s - \tau - \sigma)(cA - \lambda)} \phi(s)\psi(s)ds$$

$$- r\int_0^{t-\tau-\sigma} e^{(t - s - \tau - \sigma)cA} S(s)I(s)ds$$

(since $S(s)I(s) \geq 0$ for $0 \leq s \leq t-\tau-\sigma \leq \tau$ and $- e^{(t - s - \tau - \sigma)(-\lambda)} \geq -1$)

$$\geq e^{t(cA - \lambda)}\psi(0) + r\int_{-\tau}^{0} e^{(t - s - \tau)(cA - \lambda)} \phi(s)\psi(s)ds$$

$$+ r\int_0^{t-\tau} e^{(t - s - \tau)(cA - \lambda)} S(s)I(s)ds$$

$$- r\int_{-\tau-\sigma}^{0} e^{(t - s - \tau - \sigma)(cA - \lambda)} \phi(s)\psi(s)ds$$

$$- r(\sqrt{2c^{-1}}[V(\phi(0)) + 2\| \phi(0) \|V(\phi(0) + \psi(0))] + V(\phi(0))L^{-1})$$

(by (3.15), since $S(s)I(s) \geq 0$ for $0 \leq s \leq t-\tau-\sigma \leq \tau$)

$$\geq r\int_0^{t-\tau} e^{(t - s - \tau)(cA - \lambda)} S(s)I(s)ds \quad \text{(by (3.11))}$$

≥ 0 (since $I(s) \geq 0$ for $0 \leq s \leq t-\tau \leq \tau+\sigma$)

For $2\tau + \sigma < t \leq 3\tau + \sigma$ we repeat the argument used above for $\tau + \sigma < t \leq 2\tau + \sigma$ to conclude that $I(t) \geq 0$ for $0 \leq t \leq 3\tau + \sigma$. In fact, an induction argument yields that $I(t) \geq 0$ for all $t \geq 0$.

The following proposition provides sufficient conditions for the solutions $S(t)$, $I(t)$ of (3.6)-(3.8) to be strong solutions in the norm of X (and hence classical solutions to the problem):

PROPOSITION 3.3. Let r, τ, σ, a, c be positive constants, let λ be a nonnegative constant, let ϕ, ψ be continuous functions from $[-\tau-\sigma,0]$ to X, and let ϕ, ψ satisfy

$\phi(0)$, $\psi(0) \in D(A)$ and ϕ, ψ are Hölder continuous on $[-\tau-\sigma,0]$, that is, there exists $\gamma \in (0,1]$ and $M > 0$ such that

$$\| \phi(\theta_1) - \phi(\theta_2) \|, \quad \| \psi(\theta_1) - \psi(\theta_2) \| \leq M|\theta_1 - \theta_2|^\gamma \tag{3.16}$$

for all θ_1, $\theta_2 \in [-\tau-\sigma,0]$

Then, $S(t)$, $I(t)$ are continuously differentiable from $[0,\infty)$ to X and satisfy

$$dS(t)/dt = aAS(t) - rS(t)I(t), \quad t \geq 0 \tag{3.17}$$

$$dI(t)/dt = cAI(t) + rS(t - \tau)I(t - \tau) - rS(t - \tau - \sigma)I(t - \tau - \sigma), \quad t \geq 0 \tag{3.18}$$

$$S(t) = \phi(t), \quad I(t) = \psi(t), \quad -\tau-\sigma \leq t \leq 0 \tag{3.19}$$

Proof. We will use the fact that e^{tA}, $t \geq 0$ is an analytic semigroup in X. Equation (3.7) has the form

$$I(t) = e^{t(cA - \lambda)}\psi(0) + \int_0^t e^{(t-s)(cA-\lambda)}f(s)ds$$

where $f(s) = r[S(s - \tau)I(s - \tau) - S(s - \tau - \sigma)I(s - \tau - \sigma)]$. Since $S(t)$, $I(t)$ are Hölder continuous on $[-\tau-\sigma,0]$, $f(s)$ is Hölder continuous on $[0,\tau]$. By Theorem 1.27, p. 491 in [7] the solution $I(t)$ of (3.7) is continuously differentiable from $[0,\tau]$ to X and satisfies (3.18). Now restrict equation (3.6) to $[0,\tau]$ where it has the form as in (3.9). Since $B(s,x)$ as in (3.9) is Hölder continuous in s on $[0,\tau]$ and linear and bounded in x, we may apply Theorem 3.3.3 in [3] to conclude that $S(t)$ is continuously differentiable from $[0,\tau]$ to X and satisfies (3.17). Now use the fact that $S(t)$, $I(t)$ are Hölder continuous on $[0,\tau]$ (since they are continuously differentiable on $[0,\tau]$) and repeat the argument above to obtain that $S(t)$, $I(t)$ are continuously differentiable on $[\tau,2\tau]$ and satisfy (3.17) and (3.18) on $[\tau,2\tau]$. By repeating this argument on each step $[2\tau,3\tau]$, $[3\tau,4\tau]$,...we obtain the conclusion of the proposition.

REMARK 3.1. We note that the set of functions ϕ, ψ satisfying (3.16) is dense in $C([-\tau-\sigma,0]; X)$, where $C([-\tau-\sigma,0]; X)$ is the Banach space of continuous X-valued functions from $[-\tau-\sigma,0]$ to X with supremum norm $\|\phi\|_C = \sup_{-\tau-\sigma\leq\theta\leq 0} \|\phi(\theta)\|$, $\phi \in C([-\tau-\sigma,0]; X)$.

The proposition below demonstrates that solutions of (3.6)-(3.8) depend continuously on the initial values ϕ, ψ.

PROPOSITION 3.4. Let r, τ, σ, a, c be positive constants, let λ be a nonnegative constant, let ϕ, ψ, $\hat{\phi}$, $\hat{\psi}$ be continuous functions from $[-\tau-\sigma,0]$ to X^+, let $S(t)$, $I(t)$ and $\hat{S}(t)$, $\hat{I}(t)$ be the solutions of (3.6)-(3.8) corresponding to ϕ, ψ and $\hat{\phi}$, $\hat{\psi}$, respectively, and let $S(t)$, $I(t)$, $\hat{S}(t)$, $\hat{I}(t) \in X^+$ for all $t \geq 0$. If $L > 0$, there exists a constant $M > 0$ and an increasing function N from $[0,\infty)$ to $[0,\infty)$ such that if $\|\phi\|_C$, $\|\psi\|_C$, $\|\hat{\phi}\|_C$, $\|\hat{\psi}\|_C \leq L$, then for all $t \geq 0$

$$\|S(t) - \hat{S}(t)\| + \|I(t) - \hat{I}(t)\| \leq Me^{N(t)}[\|\phi - \hat{\phi}\|_C + \|\psi - \hat{\psi}\|_C] \tag{3.20}$$

Proof. From (3.2)-(3.7), and (3.12) we obtain that for $t \geq 0$

$$\| I(t) \| \leq \| \psi(0) \| + 2r\int_{-\tau-\sigma}^{0} \| \phi(s)\psi(s) \| ds$$
$$+ 2r\int_0^t \| I(s)S(s) \| ds \leq \| \psi(0) \| + 2r\int_{-\tau-\sigma}^{0} \| \phi(s)\psi(s) \| ds \tag{3.21}$$
$$+ 2r\| \phi(0) \| \int_0^t \| I(s) \| ds$$

Let $K = \| \psi(0) \| + 2r\int_{-\tau-\sigma}^{0} \| \phi(s)\psi(s) \| ds$ and (3.21) implies by Gronwall's Lemma that

$$\| I(t) \| \leq Ke^{2r\| \phi(0) \| t}, \qquad t \geq 0 \tag{3.22}$$

From (3.6) and (3.7) we have that

$$\| S(t) - \hat{S}(t) \| \leq \| \phi(0) - \hat{\phi}(0) \| + r\int_0^t \| S(s)I(s) - \hat{S}(s)\hat{I}(s) \| ds$$
$$+ r\int_0^t [\sup_{0\leq\xi\leq t} \| I(\xi) \| \; \| S(s) - \hat{S}(s) \| \tag{3.23}$$
$$+ \sup_{0\leq\xi\leq t} \| \hat{S}(\xi) \| \; \| I(s) - \hat{I}(s) \|] ds$$

$$\| I(t) - \hat{I}(t) \| \leq \| \psi(0) - \hat{\psi}(0) \|$$
$$+ 2r\int_{-\tau-\sigma}^{0} \| \phi(s)\psi(s) - \hat{\phi}(s)\hat{\psi}(s) \| ds$$
$$+ 2r\int_0^t [\sup_{0\leq\xi\leq t} \| I(\xi) \| \; \| S(s) - \hat{S}(s) \| \tag{3.24}$$
$$+ \sup_{0\leq\xi\leq t} \| S(\xi) \| \; \| I(s) - \hat{I}(s) \|] ds$$

Now (3.12), (3.22), (3.23), (3.24), and Gronwall's Lemma imply the conclusion of the proposition.

4. ASYMPTOTIC BEHAVIOR OF THE SOLUTIONS IN THE SPATIALLY INHOMOGENEOUS CASE

The main result we prove in this section is the following:

PROPOSITION 4.1. Let r, τ, σ, a, c be positive constants, let λ be a nonnegative constant, let ϕ, ψ be continuous functions from $[-\tau-\sigma,0]$ to X, and let the solutions $S(t)$, $I(t)$ of (3.6)-(3.8) satisfy $S(t)$, $I(t) \in X^+$ for all $t \geq 0$. The following are true:

$$\lim_{t\to\infty} S(t) \overset{\text{def}}{=} S_\infty \text{ exists (where the limit is taken in X)}$$
$$\text{and } S_\infty \text{ is a constant function on } [0,L] \tag{4.1}$$

$$\lim_{t\to\infty} I(t) \overset{\text{def}}{=} I_\infty \text{ exists (where the limit is taken in X)} \\ \text{and } I_\infty \text{ is a constant function on } [0,L] \tag{4.2}$$

$$\text{if } \lambda = 0, \text{ then } I_\infty = L^{-1}[V(\psi(0)) - r\int_{-\tau-\sigma}^{-\tau} V(\phi(s)\psi(s))ds], \\ \text{and if } I_\infty \neq 0, \text{ then } S_\infty = 0 \tag{4.3}$$

$$\text{if } \lambda > 0, \text{ then } I_\infty = 0, \text{ and if } \phi(0) > 0 \text{ then } S_\infty > 0 \tag{4.4}$$

Before we give the proof of Proposition 4.1 we prove the following lemma:

LEMMA 4.1. Let the hypothesis of Proposition 4.1 be satisfied. Then

$$\lim_{t\to\infty} \int_0^t e^{(t-s)aA}S(s)I(s)ds \overset{\text{def}}{=} I_1 \text{ exists (where the} \\ \text{limit is taken in X), } I_1 \text{ is a constant function on} \\ [0,L], \text{ and } I_1 = L^{-1}\int_0^\infty V(S(s)I(s))ds \tag{4.5}$$

$$\lim_{t\to\infty} \int_0^t e^{(t-s)(cA-\lambda)}S(s)I(s)ds \overset{\text{def}}{=} I_2 \\ \text{exists (where the limit is taken in X), } I_2 \text{ is a} \\ \text{constant function on } [0,L],\ I_2 = L^{-1}\int_0^\infty V(S(s)I(s))ds \\ \text{if } \lambda = 0 \text{ and } I_2 = 0 \text{ if } \lambda > 0 \tag{4.6}$$

Proof. It follows from (3.3) that for $T > 0$

$$\lim_{t\to\infty} \int_0^T e^{(t-s)aA}S(s)I(s)ds \overset{\text{def}}{=} K_T$$

exists, K_T is a constant function on $[0,L]$, and $K_T = \int_0^T L^{-1}V(S(s)I(s))ds$. From (3.13) we see that

$$\lim_{T\to\infty} K_T = L^{-1} \int_0^\infty V(S(s)I(s))ds \overset{\text{def}}{=} K_\infty$$

Next, we prove that $I_1 = K_\infty$. Since

$$\begin{aligned} &\| \int_0^t e^{(t-s)aA}S(s)I(s)ds - K_\infty \| \\ &\quad \leq \| K_\infty - K_T \| + \| K_T - \int_0^T e^{(t-s)aA}S(s)I(s)ds \| \\ &\quad + \| \int_0^T e^{(t-s)aA}S(s)I(s)ds - \int_0^t e^{(t-s)aA}S(s)I(s)ds \| \end{aligned}$$

it suffices to show that if $\varepsilon > 0$ there exists $M, N > 0$ such that if $T > M$ and $t > N$, then

$$\| \int_0^T e^{(t-s)aA} S(s)I(s)ds - \int_0^t e^{(t-s)aA} S(s)I(s)ds \| < \varepsilon$$

Therefore let $\varepsilon > 0$ and choose $0 < \delta < 1$ such that $\| \phi(0) \| (\pi a)^{-1/2} V(\phi(0) + \psi(0)) 2\delta^{1/2} < \varepsilon/3$. Let $M > 0$ such that $\delta^{-1/2}(\pi a)^{-1/2} \int_M^\infty V(S(s)I(s))ds < \varepsilon/3$ and $\int_M^\infty V(S(s)I(s))ds < L\varepsilon/3$. Let $N = M + 1$, let $T > M$, and let $t > N$. Then from (3.3)

$$\begin{aligned}
&\| \int_T^t e^{(t-s)aA} S(s)I(s)ds \| \\
&= \sup_{0 \le x \le L} | \int_T^t [2L^{-1} \int_0^L G(x,\xi,a(t-s))S(s)I(s)(\xi)d\xi \\
&\quad + \int_T^t L^{-1} \int_0^L S(s)I(s)(\xi)d\xi]ds | \\
&\le (\pi a)^{-1/2} \int_T^t (t-s)^{-1/2} V(S(s)I(s))ds + L^{-1} \int_T^t V(S(s)I(s))ds \\
&\le (\pi a)^{-1/2} [\sup_{T \le s \le t-\delta} (t-s)^{-1/2} \int_T^{t-\tau} V(S(s)I(s))ds \\
&\quad + \sup_{t-\delta \le s \le t} \| S(s) \| V(I(s)) \int_{t-\delta}^t (t-s)^{-1/2} ds] \\
&\quad + L^{-1} \int_T^t V(S(s)I(s))ds \\
&\le (\pi a)^{-1/2} [\delta^{-1/2} \int_M^\infty V(S(s)I(s))ds \\
&\quad + \| \phi(0) \| V(\phi(0) + \psi(0)) 2\delta^{1/2}] + L^{-1} \int_M^\infty V(S(s)I(s))ds \\
&< \varepsilon \quad \text{(where we have used (3.12) and (3.14))}
\end{aligned}$$

Thus, we have proved (4.5) and the proof of (4.6) is similar.

Proof of Proposition 4.1. The proof of (4.1) follows immediately from (3.6) and (4.5) and, from (3.3) we see that

$$S_\infty = L^{-1}(V(\phi(0)) - r\int_0^\infty V(S(s)I(s))ds) \tag{4.7}$$

The proof of (4.2) follows immediately from (3.7) and (4.6) and, from (3.3) we see that

$$\text{if } \lambda = 0,\ I_\infty = L^{-1}[V(\psi(0)) - r\int_{-\tau-\sigma}^{-\tau} V(\phi(s)\psi(s))ds] \tag{4.8}$$

$$\text{if } \lambda > 0,\ I_\infty = 0 \tag{4.9}$$

since from (3.7)

$$\begin{aligned}
I(t) &= e^{-\lambda t} e^{cAt} \psi(0) + r\int_{-\tau}^0 e^{-\lambda(t-s-\tau)} e^{(t-s-\tau)cA} S(s)I(s)ds \\
&\quad + r\int_0^{t-\tau} e^{(t-\tau-s)(cA-\lambda)} S(s)I(s)ds
\end{aligned}$$

$$- r\int_{-\tau-\sigma}^{0} e^{-\lambda(t - s - \tau - \sigma)} e^{(t - s - \tau - \sigma)cA} S(s)I(s)ds$$

$$- r\int_{0}^{t-\tau-\sigma} e^{(t - \tau - \sigma - s)(cA - \lambda)} S(s)I(s)ds$$

To establish (4.3) it remains to show that $S_\infty = 0$ if $\lambda = 0$ and $I_\infty \neq 0$. Assume for contradiction that $S_\infty > 0$. Then $I_\infty S_\infty > 0$, which implies that $\int_0^\infty V(S(s)I(s))ds = \infty$. In view of (3.13) we obtain a contradiction.

Lastly, to prove (4.4) it remains to show that if $\lambda > 0$ and $\phi(0) > 0$, then $S_\infty > 0$. Let $\varepsilon > 0$ and define

$$V_\varepsilon(\phi,\psi,t) \stackrel{\text{def}}{=} V[S(t) + I(t) - \lambda r^{-1}\ln(S(t) + \varepsilon)]$$

where $S(t)$, $I(t)$ are the solutions of (3.6)-(3.8) corresponding to ϕ, ψ. If ϕ, ψ satisfy (3.16) we may differentiate $V_\varepsilon(\phi,\psi,t)$ with respect to t to obtain

$$\begin{aligned}
\partial/\partial t\ V_\varepsilon(\phi,\psi,t) &= V(S'(t) + I'(t) - \lambda S'(t)[r(S(t) + \varepsilon)]^{-1}) \\
&= V(aAS(t) - rS(t)I(t) + cAI(t) + rS(t - \tau)I(t - \tau) \\
&\quad - rS(t - \tau - \sigma)I(t - \tau - \sigma) - \lambda I(t) \\
&\quad - \lambda(aAS(t) - rS(t)I(t))[r(S(t) + \varepsilon)]^{-1}) \\
&= V(-rS(t)I(t) + rS(t - \tau)I(t - \tau) - rS(t - \tau - \sigma)I(t - \tau - \sigma) \\
&\quad - \lambda(1 - S(t)[S(t) + \varepsilon]^{-1})I(t) - \lambda a(d/dx\ S(t))^2[r(S(t) + \varepsilon)^2]^{-1}) \\
&\leq rV(S(t - \tau)I(t - \tau))
\end{aligned}$$

which implies,

$$\begin{aligned}
V_\varepsilon(\phi,\psi,t) &\leq V_\varepsilon(\phi,\psi,0) + r\int_0^t V(S(s - \tau)I(s - \tau))ds \\
&\leq V_\varepsilon(\phi,\psi,0) + r\int_{-\tau}^\infty V(S(s)I(s))ds
\end{aligned} \tag{4.10}$$

By virtue of Remark 3.1, Proposition 3.4, and the continuity of $V_\varepsilon(\phi,\psi,t)$ from $C([-\tau-\sigma,0];\ X) \times C([-\tau-\sigma,0];\ X) \times [0,\infty)$ to R, (4.10) holds for all initial functions ϕ, ψ such that $S(t)$, $I(t) \in X^+$ for $t \geq 0$ (that is, even for those ϕ, ψ not satisfying (3.16)).

If $\phi(0) > 0$ define $M = V[\phi(0) + \psi(0) - \lambda\ln(\phi(0))r^{-1}]$. Assume for contradiction that $S_\infty = 0$. Choose $\varepsilon > 0$ such that $M + r\int_{-\tau}^\infty V(S(s)I(s))ds < -\lambda r^{-1}L\ln(2\varepsilon)$ and choose $t > 0$ such that $\|S(t)\| < \varepsilon$. Then, from (4.10)

$$\begin{aligned}
M &> V_\varepsilon(\phi,\psi,0) \\
&\geq V_\varepsilon(\phi,\psi,t) - r\int_{-\tau}^\infty V(S(s)I(s))ds \\
&\geq -\lambda r^{-1}V(\ln(S(t) + \varepsilon)) - r\int_{-\tau}^\infty V(S(s)I(s))ds
\end{aligned}$$

$$\geq -\lambda r^{-1} V(\ln(2\varepsilon)) - r\int_{-\tau}^{\infty} V(S(s)I(s))ds$$

$$= -\lambda r^{-1} L\ln(2\varepsilon) - r\int_{-\tau}^{\infty} V(S(s)I(s))ds$$

$$> M$$

Thus, we have a contradiction and so $S_\infty = 0$. The proof of Proposition 4.1 is complete.

REMARK 4.1. We conclude with some remarks concerning the diffusion coefficients a and c in (1.1) and (1.3). In [13] the results required that a = c and, as far as we know, the argument used there to demonstrate the uniform boundedness of the infective density function I(t) necessitated that a = c. The methods of Section 4 above, however, can be readily adapted to [13] in the case that a = c. In general it seems to be important to allow for unequal diffusion coefficients in a diffusion process involving subclasses of a biological population. In the diffusion of an epidemic disease within a human or animal population, for example, the diffusion coefficient c corresponding to the infective population I(t) would be very much greater than the diffusion coefficient a corresponding to the susceptible population S(t). In this case, in fact, the epidemic phenomenon might be more properly viewed as the diffusion of micro-organisms as the infective agents within a host population. In this view the infective class could be identified with the micro-organism population and the diffusion of the infective class within the population could be identified with the diffusion of the micro-organisms within the host population. The micro-organisms diffuse at a rate proportional to the gradient of the density function of the infective population from regions of high density to regions of low density. A similar identification could be made for the insect infestation of a forest or plant population. It seems, therefore, that in order for equations (1.1) and (1.3) to provide a reasonable mathematical model of a biological diffusion process such as a disease epidemic, it is necessary to allow the possibility that c is very much greater than a.

ACKNOWLEDGMENT

This work was supported in part by the National Science Foundation Grant NSF 75-06332A01.

REFERENCES

1. V. Capasso, Global solution for a diffusive nonlinear deterministic epidemic model, *SIAM J. Appl. Math.* 35 (1978), 274-284.

2. E. Conway, D. Hoff, and J. Smoller, Large time behavior of solutions of systems of nonlinear reaction-diffusion equations, *SIAM J. Appl. Math.* 35 (1978), 1-16.

3. D. Henry, *Geometric theory of semilinear parabolic equations*, to appear.

4. F. Hoppensteadt, *Mathematical theories of populations: demographics, genetics, and epidemics*, Society for Industrial and Applied Mathematics--Regional Conference Series in Applied Mathematics, Philadelphia, 1975.

5. J. Jorné and S. Carmi, Lyapunov stability of the diffusive Lotka-Volterra equations, *Mathematical Biosciences* 37 (1977), 51-61.

6. C. Kahane, On a system of nonlinear parabolic equations arising in chemical engineering, *J. Math. Anal. Appl.* 53 (1976), 343-358.

7. T. Kato, *Perturbation theory for linear operators*, Springer-Verlag, New York, 1966.

8. A. Leung, Limiting behavior for a prey-predator model with diffusion and crowding effects, *J. Math. Biology* 6 (1978), 87-93.

9. R. Martin, *Nonlinear operators and differential equations in Banach space*, Wiley-Interscience, New York, 1976.

10. J. Murray, *Lectures on nonlinear-differential-equation models in biology*, Claredon Press, Oxford, 1977.

11. H. Soper, Interpretation of periodicity in disease prevalence, *J. Royal Statist. Soc. B*, 92 (1929), 34-73.

12. P. Waltman, *Deterministic threshhold models in the theory of epidemics*, Springer-Verlag, Berlin, 1974.

13. G. Webb, A reaction-diffusion model for a deterministic diffusive epidemic, to appear.

14. S. Williams and P. -L. Chow, Nonlinear reaction-diffusion models for interacting populations, *J. Math. Anal. Appl.* 62 (1978), 157-169.

15. L. Wilson, An epidemic model involving a threshhold, *Math. Biosci.* 15 (1972), 109-121.

Part III: CONTRIBUTED PAPERS

THE CAUCHY PROBLEM FOR LINEAR FUNCTIONAL DIFFERENTIAL EQUATIONS

John A. Burns
Terry L. Herdman
Harlan W. Stech

Department of Mathematics
Virginia Polytechnic Institute and State University
Blacksburg, Virginia

1. INTRODUCTION

In this paper we consider the Cauchy problem for a large class of linear functional differential equations. In particular, we show that certain neutral functional equations (NFDE) may be formulated as dynamical systems on product spaces.

We shall restrict our attention to the finite delay case and henceforth assume that $r \in (0,+\infty)$ is fixed. Throughout the discussion we shall use L_p to denote $L_p([-r,0]; \mathbb{R}^n)$, the customary Lebesgue space of "p-integrable" functions and C to denote the usual Banach space $C([-r,0]; \mathbb{R}^n)$ of continuous functions with the sup norm. The Sobolev space of L_2 functions whose derivatives belong to L_2 will be denoted by H^1, (see [1]). If X and Y are Banach spaces, then the space of bounded linear operators from X into Y will be denoted by $B(X,Y)$. The symbol $|\cdot|$ is used to denote the norm in a Banach space, the precise space involved being clear from the context. Given a function $x : [-r,+\infty) \to \mathbb{R}^n$ and $t \geq 0$, the function $x_t : [-r,0] \to \mathbb{R}^n$ is defined by $x_t(s) = x(t + s)$.

Suppose that L is a linear $\mathbb{R}^n$-valued function with domain $D(L)$ a subspace of the Lebesgue measurable $\mathbb{R}^n$-valued functions on $[-r,0]$, and consider the retarded functional differential equation

$$\dot{x}(t) = Lx_t, \qquad t \geq 0 \tag{1.1}$$

with initial data

$$x(0) = \eta, \qquad x_o = \phi \tag{1.2}$$

where $\eta \in \mathbb{R}^n$ and $\phi \in L_2$. In 1969, Borisovič and Turbabin [2] announced the following result.

THEOREM 1.1. Suppose that a linear function L satisfies the three conditions:

(α) $C \subseteq D(L)$ with L restricted to C a bounded operator,

(β) for each $t_1 > 0$ and $x \in L_2(-r,t_1)$, $t \to Lx_t$ defines a measurable function on $[0,t_1]$ and this function depends only on the equivalence class of x,

(γ) the function $g(t) \equiv Lx_t$ belongs to $L_1(0,t_1)$ for each $t_1 \geq 0$ and there is a continuous function Γ such that

$$\int_0^t |Lx_s|ds \leq \Gamma(t)[\int_{-r}^t |x(s)|^2 ds]^{1/2}$$

for all $t \in [0,t_1]$.

If $(\eta,\phi) \in \mathbb{R}^n \times L_2$, then there exists a unique function $x : [-r,+\infty) \to \mathbb{R}^n$ such that x is absolutely continuous (A.C.) on $[0,t]$ for each $t \geq 0$, $x(0) = \eta$, $x(s) = \phi(s)$ a.e. on $[-r,0]$ and x satisfies (1.1) a.e. on $[0,+\infty)$. Moreover, if $S(t) : \mathbb{R}^n \times L_2 \to \mathbb{R}^n \times L_2$ is defined by

$$S(t)\ (\eta,\phi) = (x(t),x_t) \tag{1.3}$$

then $\{S(t)\}_{t\geq 0}$ forms a C_o-semigroup on $\mathbb{R}^n \times L_2$ with infinitesimal generator A defined by

$$D(A) = \{(\eta,\phi)|\phi \in H^1,\ \phi(0) = \eta\} \tag{1.4}$$

and

$$A(\eta,\phi) = (L\phi,\dot{\phi}) \tag{1.5}$$

In 1977, Vinter [11] observed that in some sense the three conditions imposed by Borisovič and Turbabin were redundant. Vinter defined a linear operator L to be C-bounded if $H^1 \subseteq \mathcal{D}(L) \subseteq C$ and there exists a monotone increasing function $\nu : [-r,0] \to \mathbb{R}$ such that

$$|L\phi| \leq \int_{-r}^{0} |\phi(s)|d\nu(s)$$

for all $\phi \in \mathcal{D}(L)$. Using only the condition that L be C-bounded, Vinter showed that the operator A (defined by (1.4)-(1.5)) is the infinitesimal generator of a C_o-semigroup on $\mathbb{R}^n \times L_2$. In that same paper, Vinter raised the question of whether the above result is true for L belonging to $\mathcal{B}(H^1, \mathbb{R}^n)$.

In 1978, Delfour [4] proved a slightly modified version of this result using only the assumption that $L \in \mathcal{B}(H^1, \mathbb{R}^n)$ and hence provided an answer to Vinter's question. More precisely, Delfour showed that if $L \in \mathcal{B}(H^1, \mathbb{R}^n)$ then the operator A (again defined by (1.4)-(1.5)) is the infinitesimal generator of a C_o-semigroup on $\mathbb{R}^n \times L_2$. Moreover the semigroup is of the form (1.3). In the same paper, Delfour stated that the converse was also true: if A generates a C_o-semigroup on $\mathbb{R}^n \times L_2$, then it is necessary that $L \in \mathcal{B}(H^1, \mathbb{R}^n)$.

The purpose of this paper is to extend the aforementioned results to a class of neutral functional differential equations. In Section 2 we state and prove a necessary condition that (for finite delay systems) includes as a special case the necessary condition announced by Delfour. Sufficient conditions are stated in Section 3 and we conclude with additional comments concerning the literature.

2. SEMIGROUPS ON PRODUCT SPACES AND NEUTRAL FUNCTIONAL DIFFERENTIAL EQUATIONS

Let L and D be linear $\mathbb{R}^n$-valued functions with domains $\mathcal{D}(L)$ and $\mathcal{D}(D)$ subspaces of the Lebesgue measurable $\mathbb{R}^n$-valued functions on [-r,0]. For the remainder of the paper we assume that H^1 is contained in both $\mathcal{D}(L)$ and $\mathcal{D}(D)$. However, at this point no continuity assumptions are made on L or D. Define the operator A on

$$\mathcal{D}(A) \equiv \{(\eta,\phi) \in \mathbb{R}^n \times L_2 | \phi \in H^1, D\phi = \eta\} \tag{2.1}$$

by

$$A(\eta,\phi) = (L\phi,\dot{\phi}) \tag{2.2}$$

REMARK 2.1. In the special case when the operator D is defined on $\mathcal{D}(D) = H^1$ by $D\phi = \phi(0)$, the operator reduces to the operator defined by (1.4)-(1.5).

In this section we investigate some conditions that are necessary of the operators L and D if A (defined by (2.1)-(2.2)) is the infinitesimal generator of a C_o-semigroup on $\mathbb{R}^n \times L_2$. The following lemma will be useful.

LEMMA 2.2. If for $\lambda \in \mathbb{C}$ and $\psi \in L_2$ the function $M_\lambda\psi : [-r,0] \to \mathbb{R}^n$ is defined by

$$[M_\lambda\psi](s) = \int_0^s e^{\lambda(s - u)}\psi(u)du$$

then M_λ defines an isomorphism of L_2 onto $H_0^1 \equiv \{\phi \in H^1 | \phi(0) = 0\}$.

Proof. Elements of H^1 are continuous and the natural injection $H^1 \subseteq C$ is continuous (see [1]). Thus H_0^1 is a closed subspace of H^1. Since $d/ds\ [M_\lambda\psi](s) = \lambda[M_\lambda\psi](s) + \psi(s)$ a.e. and $[M_\lambda\psi](0) = 0$, M_λ maps L_2 into H_0^1. If $\phi \in H_0^1$, then $[\dot{\phi} - \lambda\phi] \in L_2$ and $\phi = M_\lambda[\dot{\phi} - \lambda\phi]$. Consequently, M_λ has range equal to H_0^1. It is well known that M_λ defines a bounded linear operator on L_2. If $\psi_n \in L_2$, $n = 1,2,\ldots,$ and $\psi_n \to \psi$, then

$$\frac{d}{ds}\,[M_\lambda\psi_n] = \lambda[M_\lambda\psi_n] + \psi_n \to \lambda[M_\lambda\psi] + \psi = \frac{d}{ds}\,[M_\lambda\psi]$$

Continuity of M_λ as an operator from L_2 onto H^1 follows immediately as does the continuity of M_λ^{-1}.

THEOREM 2.3. If A defined by (2.1)-(2.2) is the infinitesimal generator of a C_o-semigroup on $\mathbb{R}^n \times L_2$, then both L and D belong to $\mathcal{B}(H^1, \mathbb{R}^n)$.

Proof. Assume A generates a C_o-semigroup on $\mathbb{R}^n \times L_2$. The Hille-Yosida Theorem (see [8]) implies that A is densely defined, closed and there exists a real ω such that if $\lambda > \omega$ then λ belongs to $\rho(A)$, the resolvent set of A. The existence of such an ω implies that for $\lambda > \omega$ and $(\xi,\psi) \in \mathbb{R}^n \times L_2$ there is a unique solution $(\eta,\phi) \in D(A)$ to the equation

$$(A - \lambda I)(\eta,\phi) = (\xi,\psi) \tag{2.3}$$

and this solution depends continuously on (ξ,ψ). Considering the definition of A, equation (2.3) is equivalent to

$$L\phi - \lambda\eta = \xi \tag{2.4}$$

$$\dot{\phi} - \lambda\phi = \psi \tag{2.5}$$

and

$$D\phi = \eta \tag{2.6}$$

where $\phi \in H^1$. Equation (2.5) implies that

$$\phi(s) = e^{\lambda s}c + [M_\lambda\psi](s) \tag{2.7}$$

for some $c \in \mathbb{R}^n$. Here c must be taken so that (2.4) and (2.6) hold. Substituting (2.7) into both (2.4), (2.6) and eliminating η we obtain

$$\Delta(\lambda)c = -\xi - \lambda D[M_\lambda\psi] + L[M_\lambda\psi] \tag{2.8}$$

where $\Delta(\lambda) \equiv \lambda D(e^{\lambda\cdot}I) - L(e^{\lambda\cdot}I)$. For $\lambda > \omega$ equation (2.8) must be uniquely solvable for c. Thus, $\Delta(\lambda)$ must be nonsingular for $\lambda > \omega$. From (2.7), (2.8) and the continuous dependence of ϕ on (ξ,ψ) we see that for $\lambda > \omega$ the $\mathbb{R}^n$-valued operator $(\lambda D - L)\,M_\lambda$ must belong to $B(L_2, \mathbb{R}^n)$. Lemma 2.2 implies that $(\lambda D - L) = [(\lambda D - L)\,M_\lambda]\,M_\lambda^{-1}$ belongs to $B(H_0^1, \mathbb{R}^n)$. Therefore, $(\lambda D - L)$ is bounded on H^1 into $\mathbb{R}^n$ for all $\lambda > \omega$ which clearly implies that D and L belong to $B(H^1, \mathbb{R}^n)$.

If A generates a C_0-semigroup on $R^n \times L_2$ we can apply Theorem 2.3 together with standard representation theorems (see [1]) to obtain $n \times n$ matrix valued functions A, B, F and G, whose column vectors belong to L_2, such that if $\phi \in H^1$, then

$$L\phi = \int_{-r}^{0} (F(s)\phi(s) + G(s)\dot{\phi}(s))ds \tag{2.9}$$

and

$$D\phi = \int_{-r}^{0} (A(s)\phi(s) + B(s)\dot{\phi}(s))ds \tag{2.10}$$

The representation (2.9) will be used in the next theorem.

REMARK 2.4. It is easy to show that if A generates a C_o-semigroup, then in the representation (2.10) the matrix function B can not be identically zero.

THEOREM 2.5. Assume that A generates a C_o-semigroup $\{S(t)\}_{t\geq 0}$. Let L have the representation (2.9). If $(\eta,\phi) \in \mathbb{R}^n \times L_2$, then there is a unique function $x : [-r,+\infty) \to \mathbb{R}^n$ such that $x_t \in L_2$ for all $t \geq 0$, $x_o = \phi$ and

$$S(t)\ (\eta,\phi) = (y(t),\ x_t) \tag{2.11}$$

where

$$\begin{aligned} y(t) &= \eta + \int_{-r}^{0} G(u)\ (x(t+u) - \phi(u))ds \\ &\quad + \int_0^t \int_{-r}^{0} F(u)x(s+u)duds \end{aligned} \tag{2.12}$$

Moreover, if $(\eta,\phi) \in D(A)$ then for each $t \geq 0$, $x_t \in H^1$, Dx_t is continuously differentiable and

$$S(t)\ (\eta,\phi) = (Dx_t,\ x_t) \tag{2.13}$$

In particular, x is the unique H^1 solution to the NFDE

$$\frac{d}{dt} Dx_t = Lx_t \tag{2.14}$$

with initial data

$$x_o = \phi \tag{2.15}$$

Proof. Assume that $(\eta,\phi) \in D(A)$. If A generates a C_o-semigroup $\{S(t)\}_{t\geq 0}$, then the Cauchy problem

$$\dot{z}(t) = Az(t), \qquad z(0) = (\eta,\phi) \in D(A) \tag{2.16}$$

is uniformly correct (see [5]). Moreover, $z(t) = (z_1(t), z_2(t,\cdot)) = S(t)(\eta,\phi)$ is the unique continuously differentiable solution to (2.16). For each $t \geq 0$, $z(t) \in \mathcal{D}(A)$ so that $z_2(t,\cdot) \in H^1$. The Cauchy problem (2.16) may be written as

$$\frac{d}{dt} z_1(t) = Lz_2(t,\cdot), \qquad z_1(0) = \eta \tag{2.17}$$

$$\frac{\partial}{\partial t} z_2(t,s) = \frac{\partial}{\partial s} z_2(t,s), \qquad z_2(0,s) = \phi(s) \tag{2.18}$$

Equation (2.18) implies that there exists a function $x : [-r,+\infty) \to \mathbb{R}^n$ such that x is absolutely continuous on each interval $[-r,t)$, $x_o = \phi$ and $z_2(t,s) = x(t + s)$ for $t \geq 0$ and $s \in [-r,0]$. Moreover, $(z_1(t), z_2(t,\cdot)) = (z_1(t), x_t) \in \mathcal{D}(A)$ implies that $z_1(t) = Dx_t$. Consequently, we have shown that $S(t)(\eta,\phi) = (Dx_t, x_t)$, $x_t \in H^1$, $Dx_t = z_1(t)$ is continuously differentiable and x satisfies the NFDE (2.14) with $x_o = \phi$. If $(\eta,\phi) \in \mathcal{D}(A)$, then we may integrate (2.14) to obtain

$$\begin{aligned} Dx_t &= D\phi + \int_0^t Lx_s ds \\ &= \eta + \int_0^t \int_{-r}^0 (F(u)x(s + u) + G(u)\dot{x}(s + u))duds \\ &= \eta + \int_0^t \int_{-r}^0 F(u)x(s + u)duds + \int_{-r}^0 \int_0^t G(u)\dot{x}(s + u)dsdu \end{aligned}$$

Consequently, we have that

$$\begin{aligned} Dx_t = \eta &+ \int_0^t \int_{-r}^0 F(u)x(s + u)duds \\ &+ \int_{-r}^0 G(u)(x(t + u) - x(u))du \end{aligned} \tag{2.19}$$

for $(\eta,\phi) \in \mathcal{D}(A)$.

Consider now an arbitrary $(\eta,\phi) \in \mathbb{R}^n \times L_2$. Let $(\eta^m,\phi^m) \in \mathcal{D}(A)$, $m = 1, 2, 3, \ldots$, be such that $(\eta^m,\phi^m) \to (\eta,\phi)$ and let x^m be the corresponding solution to the NFDE (2.14). Strong continuity of S(t) implies that

$$\lim_{m\to\infty} (Dx_t^m, x_t^m) = \lim_{m\to\infty} S(t)(\eta^m,\phi^m) = S(t)(\eta,\phi) \equiv (y_1(t), y_2(t,\cdot))$$

for each $t \geq 0$. In particular, x_t^m is a Cauchy sequence in L_2 for each $t \geq 0$. This readily implies that x^m is a Cauchy sequence in $L_2(-r,t)$ for all $t \geq 0$. Therefore, there is a unique function $x : [-r,+\infty) \to \mathbb{R}^n$ such that for each $t \geq 0$, $x \in L_2(-r,t)$, $x_t^m \to x_t$ and

$$\lim_{m\to\infty} \int_{-r}^{t} |x^m(s) - x(s)|^2 ds = 0$$

Since $x_t^m \to y_2(t,\cdot)$, we have that $y_2(t,\cdot) = x_t$. Let $y(t)$ be defined by (2.12) and use (2.19) to obtain

$$\begin{aligned} |y(t) - y_1(t)| &\leq |y(t) - Dx_t^m| + |Dx_t^m - y_1(t)| \\ &\leq |Dx_t^m - y_1(t)| + |\eta - \eta^m| + \int_{-r}^{0} |G(u)|\ |x(t + u) - x^m(t + u)|du \\ &\quad + \int_{-r}^{0} |G(u)|\ |\phi(u) - \phi^m(u)|du + \int_0^t \int_{-r}^{0} |F(u)|\ |x(s + u) \\ &\qquad - x^m(s + u)|duds \end{aligned}$$

We apply the Cauchy-Schwarz inequality along with the convolution theorem to the above inequality to obtain (see [3] where similar inequalities are given)

$$\begin{aligned} |y(t) - y_1(t)| &\leq |Dx_t^m - y_1(t)| + |\eta - \eta^m| \\ &\quad + \left(\int_{-r}^{0} |G(u)|^2du\right)^{1/2} (|x_t - x_t^m| + |\phi - \phi^m|) \\ &\quad + (t)^{1/2} \left(\int_{-r}^{0} |F(u)|^2du\right)^{1/2} \left(\int_{-r}^{t} |x(u) - x^m(u)|^2du\right)^{1/2} \end{aligned}$$

Each term on the right side of inequality (2.20) goes to zero as $m \to +\infty$, proving that $y(t) = y_1(t)$. Thus we have shown that $S(t)(\eta,\phi) = (y_1(t), y_2(t,\cdot)) = (y(t),x_t)$.

3. GENERALIZED SOLUTIONS OF NEUTRAL FUNCTIONAL DIFFERENTIAL EQUATIONS

In this section we give conditions on L and D that are sufficient for A to generate a C_o-semigroup on $\mathbb{R}^n \times L_2$. If D is a bounded linear operator from C into $\mathbb{R}^n$, then there is a matrix valued function $\mu : [-r,0] \to \mathbb{R}^{n\times n}$ whose entries are of bounded variation, continuous from the right on $(-r,0)$, $\mu(-r) = 0$ and for each $\phi \in C$

$$D\phi = \int_{-r}^{0} [d\mu(s)]\phi(s) \tag{3.1}$$

Such an operator D is said to be *atomic* at $s \in [-r,0]$ if the jump $\mu(s) - \mu(s^-)$ is nonsingular.

THEOREM 3.1. If $L \in \mathcal{B}(H^1, \mathbb{R}^n)$ and $D \in \mathcal{B}(C, \mathbb{R}^n)$, with D atomic at zero, then the operator $A(\eta,\phi) \equiv (L\phi,\dot{\phi})$ defined on $\mathcal{D}(A) = \{(\eta,\phi) \mid \phi \in H^1, D\phi = \eta\}$ is the infinitesimal generator of a C_o-semigroup $\{S(t)\}_{t\geq 0}$ on $\mathbb{R}^n \times L_2$.

The proof of Theorem 3.1 requires a number of technical preliminaries and will therefore appear in a forthcoming paper.

Theorem 3.1 can be considered as an extension of the results of Hale [6] and Henry [7] concerning the existence of solutions for the NFDE (2.14)-(2.15). In particular, given any initial function $\phi \in L_2$ and "starting value" $\eta \in \mathbb{R}^n$ we define the generalized solution of (2.14)-(2.15) to be the function x defined in Theorem 2.5. In Hale [6] it is shown that if D and L belong to $\mathcal{B}(C, \mathbb{R}^n)$, D atomic at zero and $\phi \in C$, then there is a unique continuous function $w = w(\cdot,\phi) : [-r,+\infty) \to \mathbb{R}^n$ for which $w_o = \phi$, Dw_t is continuously differentiable and satisfies $d/dt\, Dw_t = Lw_t$. The usual solution operator $T(t)\phi = w_t(\cdot,\phi)$, $t \geq 0$ defines a C_o-semigroup on C. The next theorem shows that the semigroup $\{S(t)\}_{t\geq 0}$ generated by A "extends" the semigroup $\{T(t)\}_{t\geq 0}$.

THEOREM 3.2. Assume D and L belong to $\mathcal{B}(C, \mathbb{R}^n)$ and D is atomic at zero. If S(t) and T(t) are as above, then

$$S(t)(D\phi,\phi) = (DT(t)\phi,\ T(t)\phi)$$

for all $t \geq 0$, and each $\phi \in C$.

Proof. Since H^1 is a dense subset of C we may take a sequence $\{\phi^m\}$ in H^1 with $\phi^m \to \phi$ uniformly on $[-r,0]$. As in [6], we have that (2.14)-(2.15) (with $\phi = \phi^m$) has a unique continuous solution

$w^m = w^m(\cdot,\phi^m) : [-r,+\infty) \to \mathbb{R}^n$. Theorem 3.1, Theorem 2.5 and uniqueness imply that $w_t(\cdot,\phi^m) = T(t)\phi^m$ is the L_2 coordinate of $S(t)\ (D\phi^m,\phi^m)$. Therefore, we have that

$$\lim_{m\to\infty} S(t)\ (D\phi^m,\phi^m) = \lim_{m\to\infty} (DT(t)\phi^m,\ T(t)\phi^m) = (DT(t)\phi,\ T(t)\phi)$$

in $\mathbb{R}^n \times C$. This implies that in the product space $\mathbb{R}^n \times L_2$,

$$S(t)(D\phi,\phi) = \lim_{m\to\infty} S(t)(D\phi^m,\phi^m) = (DT(t)\phi,\ T(t)\phi)$$

as desired.

The results presented above are closely related to the recent work of Kunish [9] where similar questions are considered within the product space $\mathbb{R}^n \times L_p([-r,0];\mathbb{R}^n)$. Marcus and Mizel [10] also consider the well posedness of certain functional partial differential equations on product spaces. The hypotheses of Marcus and Mizel are described in terms of the "operators" defined by the equation in lieu of conditions on the "functionals" L and D. In addition to product spaces other phase spaces have been used (see Hale [6] and the numerous references therein).

ACKNOWLEDGMENT

The first author was supported in part by the Army Research Office under grant DAAG-29-78-G-0125.

REFERENCES

1. R. A. Adams, *Sobolev Spaces*, Academic Press, New York, 1975.

2. Ju. Borisovič and A. S. Turbabin, On the Cauchy problem for linear nonhomogeneous differential equations with retarded arguments. *Soviet Math. Dokl.*, 10 (1969), pp. 401-405.

3. J. A. Burns and T. L. Herdman, Adjoint semigroup theory for a class of functional differential equations, *SIAM J. Math. Anal.*, 5 (1976), pp. 729-745.

4. M. C. Delfour, The largest class of hereditary systems defining a C_o-semigroup on the product space, *Canadian J. of Math.*, to appear.

5. S. G. Krein, *Linear Differential Equations in Banach Space*, Translations, Amer. Math. Soc. 29, Providence, R.I., 1971.

6. J. K. Hale, *Theory of Functional Differential Equations*, Springer-Verlag, New York, 1977.

7. D. Henry, Linear autonomous functional differential equations of neutral type in Sobolev space $W_2^{(1)}$, unpublished manuscript.

8. E. Hille and R. S. Phillips, *Functional Analysis and Semi-Groups*, Waverly Press, Baltimore, Md., 1957.

9. K. Kunisch, Neutral functional-differential equations in L^p-spaces and averaging approximations, *J. of Nonlinear Analysis*, to appear.

10. M. Marcus and V. J. Mizel, Nonlocal boundary problems for functional partial differential equations, *Integral and Functional Differential Equations*, Marcel Dekker, New York, 1981 (this volume).

11. R. B. Vinter, On a problem of Zabczyk concerning semigroups generated by operators with non-local boundary conditions, Publication 7718, (1977), Department of Computing and Control, Imperial College of Science and Technology, London (England).

INTEGRAL EQUATIONS ARISING IN SOME N-PERSON GAMES AND THEIR ASYMPTOTIC SOLUTION

Harvey Diamond

Department of Mathematics
West Virginia University
Morgantown, West Virginia

1. INTRODUCTION

In this paper we analyze two symmetric, noncooperative N-person games. The object of our analysis is the calculation of the symmetric equilibrium, especially when N is large. The equations which determine the symmetric equilibrium turn out to be integral equations. Using techniques from singular perturbation theory, we are able to obtain asymptotic approximations to the solutions of the integral equations when N is large. This is a basic outline of the paper; further explanation and elaboration will appear in the body of the paper. We begin then with our first example.

EXAMPLE 1. In this game N players each independently choose a $t \in [a,b]$. Let $\{t_i\}_{i=1}^N$ be those choices, ordered so that $a \equiv t_o < t_1 < \ldots < t_N$. The player choosing t_i receives a payoff $G(t_i) - G(t_{i-1})$ where the function G is continuous, monotone increasing, and bounded.

The definition of this game can be motivated with some "real life" situations. Consider first N theives, each of whom possesses a master key to some parking meters. At $t_o = 0$ a legitimate collector removes all the money in the meters; subsequently money is deposited in the meters so that the total deposited between t_o and time t is G(t) (G(t) may alternately represent an expected value of money deposited). The theives must independently decide when to come and remove the money in the meters. Thus the thief arriving at time t_i (say) will obtain the amount of money deposited since the previous thief removed all the money at time t_{i-1}, namely $G(t_i) - G(t_{i-1})$. The game continues until the next legitimate collection at time t = b. All parameters and functions are known by the players.

Another application of Example 1 is The Price is Right. In this game N players independently try to guess the actual price of an object, given a probability distribution for its price, $F_p(t) = \Pr(\text{Price} \leq t)$. The player guessing closest to the actual price without going over wins, with say, a payoff of 1. If $\{t_i\}_1^N$ represent the ordered price guesses of the players then the player choosing t_i has as his expected payoff precisely his probability of winning, which is the probability that the actual price falls between t_i and t_{i+1}, namely $F_p(t_{i+1}) - F_p(t_i)$. Reversing the direction of positive t puts this payoff function into the form of Example 1.

In noncooperative games each player seeks to maximize his own payoff. The players are permitted to make their choices randomly, i.e. according to some probability distribution, and each player then seeks to maximize his expected payoff.

One solution concept for noncooperative games is the equilibrium. An equilibrium is a set of randomized strategies for the players with the property that no player can improve his expected payoff by unilaterally deviating from his assigned equilibrium strategy. In general, as a prescription for a player's actions in a noncooperative game, the equilibrium solution concept is not satisfactory because more than one equilibrium may exist with different equilibrium payoffs. However, in symmetric games such as ours, where the players are identical, the *symmetric equilibrium* has some intuitive appeal as an "optimal" strategy for the players and hence will be the focus of our attention for the remainder of the paper. The symmetric equilibrium will be a probability distribution which produces an equilibrium when used by all the players in making their choices. We proceed then to develop the equations which determine this equilibrium probability distribution.

2. FORMULATION OF EQUATIONS FOR EQUILIBRIUM

In this section we develop the integral equations which determine the symmetric equilibrium for Example 1. Our first observation is that, without loss of generality we may take $[a,b] = [0,1]$ and $G(t) = t$. For, suppose $F(t)$ is the equilibrium probability distribution for $G(t) = t$ and consider, say, $G(t) = \overline{G}(t)$. If $F(\cdot)$ is used by each player to randomly pick a *percentile* of $\overline{G}(t)$, the player then choosing the value of t corresponding to that percentile, then the strategy so defined is an equilibrium strategy for the game with $G = \overline{G}$. More formally, if each player chooses a t in the game $G = \overline{G}$ according to the distribution $\Pr\{(G(t) - G(a))/(G(b) - G(a)) \leq s\} = F(s)$, an equilibrium is obtained. We will henceforth consider $[a,b] = [0,1]$ and $G(t) = t$ to hold.

Assume that N-1 players use the equilibrium distribution $F(t)$ and suppose the remaining player chooses $t = s$. Let the random variable τ denote his payoff. We then have

$$\Pr\{\tau > r\} = \begin{cases} [1 - F(s) + F(s - r)]^{N-1} & 0 \leq r < s \\ 0 & r \geq s \end{cases} \tag{1}$$

This is true because $\tau > r$ holds if and only if all of the other choices fall outside the interval $[s - r, s]$; and this occurs with zero probability if $r \geq s$ and the probability given above if $0 \leq r < s$, since the $N - 1$ other choices are independently made using the distribution F.

The expected payoff at $t = s$, denoted $C(s)$, is then given by

$$C(s) = E\{\tau\} = \int_0^\infty \Pr\{\tau > r\}dr = \int_0^s [1 - F(s) + F(s - r)]^{N-1}dr \tag{2}$$

The next step is to determine and/or guess the structure of F, that is, the general location of intervals where $F'(t) = 0$, $F'(t) > 0$, and points (if any) where $F(t)$ is discontinuous. We note for our game that choosing arbitrarily small values of t cannot be optimal, for these give arbitrarily small payoff. Thus we will assume that $F(t)$ has the structure:

$$F(t) = 0 \quad \text{for} \quad 0 \le t \le c; \; F'(t) > 0 \quad \text{for} \quad c \le t \le 1; \quad F(t) \text{ continuous} \tag{3}$$

The constant c is to be determined. The assumed structure for F will be justified when we find such an F satisfying the conditions of an equilibrium.

Using (3) we can rewrite the payoff function C(s) from (2) as

$$C(s) = \begin{cases} c[1 - F(s)]^{N-1} + \int_c^s [1 - F(s) + F(r)]^{N-1} dr, & s \ge c \\ s, \quad s \le c \end{cases} \tag{4}$$

If F(t) is an equilibrium strategy, it must be an optimal strategy for the remaining player, whose payoff is given by C(s) when he chooses $t = s$. But F will be optimal if and only if $C(s) = C(c) = c$ where $F'(s) > 0$, namely $c \le s \le 1$; and $C(s) \le c$ where $F'(s) = 0$, namely $0 \le s \le c$. The latter requirement is clearly satisfied: if his opponents never choose a $t < c$, it cannot be optimal for the remaining player to do so; this is reflected in (4). The former requirement leads to the desired integral equation for F(t):

$$1 = [1 - F(s)]^{N-1} + (1/c) \int_c^s [1 - F(s) + F(r)]^{N-1} dr, \qquad s \in [c,1] \tag{5a}$$

with additional requirements on F:

$$F(c) = 0, \qquad F(1) = 1 \tag{5b}$$

3. ANALYSIS OF THE INTEGRAL EQUATION

The nonlinear integral equation (5a) can be transformed via

$$x = F(s), \qquad y = F(r), \qquad h(y) = \frac{1}{c}\frac{d}{dy} F^{-1}(y) \tag{6}$$

into the linear integral equation for h(y) given by

$$1 = (1 - x)^{N-1} + \int_0^x (1 - x + y)^{N-1} h(y)dy, \qquad x \in [0,1] \tag{7}$$

Given h(y), F(t) and c are found using (6) and (5b).

The integral equation (7) is, in principle, easy to solve. One can use a variety of numerical methods or the Laplace Transform method. Analytic solutions for N = 2 and N = 3 are given by

$$N = 2: \quad h(x) = e^x; \; F(x) = \ln(ex), \qquad x \in [e^{-1},1], \qquad c = e^{-1}$$

$$N = 3: \quad h(x) = 2e^x \cos(x); \; F^{-1}(x) = ce^x[\cos(x) + \sin(x)];$$

$$c = e[\cos(1) + \sin(1)]^{-1} \simeq .266$$

For N larger than three, the Laplace Transform of h(x) is given by

$$\mathcal{L}(h) = \frac{\left[\frac{p^{n-1}}{(n-1)!} - \frac{p^{n-2}}{(n-2)!} + \ldots + (-1)^{n-1}\right]}{\left[\frac{p^n}{n!} - \frac{p^{n-1}}{(n-1)!} + \ldots + (-1)^n\right]}, \qquad n = N - 1$$

We see that h(x) is a linear combination of exponentials, whose exponents are the roots of $\sum_{k=0}^{n} (-1)^k p^k/k!$. As N increases, it becomes more difficult to determine these roots, at most one of which is real. (This last assertion is known but not obvious.) Further, such a solution gives no useful information about the behavior of h(x), especially when N is large. In the next section we will obtain asymptotic approximations to h(x) which do furnish such information.

For future reference, we differentiate (7), obtaining

$$(1 - x)^{N-2} = \frac{h(x)}{N-1} - \int_0^x (1 - x + y)^{N-2} h(y)dy, \qquad x \in [0,1] \tag{8}$$

whence we obtain h(0) = N - 1. Equation (8) can be used to generate numerical solutions to h(x).

4. ASYMPTOTIC ANALYSIS

As stated before, standard methods of solving (7) become cumbersome as N increases; they do not give useful information about the behavior of h;

their accuracy decreases. In this section we will use some techniques of asymptotic analysis, particularly singular perturbation theory, to develop analytic approximations to h(x) when N is large.

Part of the asymptotic behavior of h(x) can be motivated by considering a game related to Example 1. To obtain this game we take the interval [0,1] in Example 1 and connect 0 and 1 together to form a circle, the connection being located at the bottom of the circle. The payoff to a player in this game is then defined as the distance to the closest choice in the counter-clockwise direction--the connection no longer has any significance. It is easy to verify that the uniform distribution on the circle constitutes a symmetric equilibrium for this game, with equilibrium payoff 1/N.

Example 1 can be obtained from this game if we include an $N + 1^{st}$ dummy player who always chooses the point formed by the connection of 0 and 1. It is plausible that when N is large the dummy player has little effect on the equilibrium. Thus, roughly speaking, we expect $F(t) \simeq t$, $c \simeq 1/N$ and $h(x) \simeq N$ to hold approximately in some sense when N is large. With this in mind we proceed to the asymptotic analysis.

We rewrite (7) in operator form as

$$L[h(x)] = \int_0^x (1 - x + y)^{N-1} h(y)dy + (1 - x)^{N-1} - 1 = 0 \tag{9}$$

We want to develop a sequence of functions, asymptotic approximations to h(x), which satisfy (9) to $O(1/N^k)$, k = 0, 1,..., uniformly in x for $x \in [0,1]$.

A. Outer Expansion

If we try the approximation $h(x) \sim N$ in (9), we obtain $L[N] = (1 - x)^{N-1} - (1 - x)^N$. We observe that L[N] is exponentially small in the region x = O(1), i.e. for any fixed x bounded away from zero. Where x = O(1/N), L[N] is no longer exponentially small and the approximation $h(x) \sim N$ breaks down. In the language of singular perturbation theory, $h(x) \sim N$ is called the *outer expansion* for h. We write

$$h(x) = N + \text{e.s.t.}, \qquad x = O(1) \tag{10}$$

where e.s.t. stands for "exponentially small terms."

B. Inner Expansion

In the region $x = O(1/N)$ a different approximation for $h(x)$ must be generated. This region is called a *boundary layer*. It is a region of rapid change of $h(x)$ (or its derivatives), reflecting the rapid change of $(1 - x)^N$ in the same region. (For instance, $h(0) = N - 1$ while $h(x) = N$ + e.s.t. for $x = O(1)$ implying $h'(x) = O(N)$ for $x = O(1/N)$.) The first step in the procedure is to rescale the variables x and y to reflect their true size of $O(1/N)$ by introducing the *stretched* variables u and v given by $x = u/N$, $y = v/N$ and the new function $h^*(u) = h(x)$. By doing this we will be able to work with variables u and v of size $O(1)$ and a function h^* which varies smoothly with u and v. The next step is to expand, in powers of $1/N$, the functions which appear in (9):

$$(1 - x)^{N-1} = (1 - u/N)^{N-1} \sim e^{-u} - \frac{(u^2 - 2u)}{2N} e^{-u} + \frac{(3u^4 - 20u^3 + 24u^2)}{24N^2} e^{-u} + \ldots \tag{11}$$

$$(1 - x + y)^{N-1} = \left(1 - \frac{u - v}{N}\right)^{N-1} \sim \text{(same as above with } u - v \text{ in place of } u)$$

We assume for $h^*(u)$ an asymptotic expansion of the form

$$h^*(u) \sim h_0^*(u) + h_1^*(u)/N + h_2^*(u)/N^2 + \ldots \tag{12}$$

If we plug (11) and (12) into (9) and demand that the coefficients of $1/N^k$, $k = 0, 1, \ldots$ in (9) vanish, we obtain a sequence of integral equations for the $h_i^*(u)$ in (12). These turn out to be simple linear integral equations which can be solved by differentiation; the $h_i^*(u)$ are solved for recursively. The resulting three term expansion for $h^*(u)$ is given by

$$h^*(u) \sim N[1 - e^{-u}/N + (u^2 - u)e^{-u}/N^2 + \ldots] \tag{13}$$

The inner expansion is given by $h(x) = h^*(Nx)$ and is valid for $x = O(1/N)$. Using the first three terms in (13) gives $L[h^*(u)] = O(1/N^3)$.

We observe that when $x = 0(1)$

$$h^*(Nx) \sim N[1 - e^{-Nx}/N + (N^2x^2 - Nx)e^{-Nx}/N^2 + \ldots]$$
$$= N + \text{e.s.t.} \sim h(x), \qquad x = 0(1)$$

Thus $h^*(Nx)$, the inner solution, matches up with $h(x)$, the outer solution, with exponentially small error when $x = 0(1)$; using the three term expansion from (13) we have

$$L[h^*(Nx)] = 0(1/N^3) \qquad \text{uniformly for} \quad 0 \leq x \leq 1 \tag{14}$$

so that the three term expansion of $h^*(Nx)$ satisfies (9) uniformly to $0(1/N^3)$ and furnishes the asymptotic approximation

$$h(x) \sim N[1 - e^{-Nx}/N + (N^2x^2 - Nx)e^{-Nx}/N^2] \tag{15}$$

which will henceforth be used.

Extensive numerical verification of (15) has not been carried out. A comparison of the analytic solution of $h(x)$ with the asymptotic approximation (15) for $N = 4$ shows an error less than .05 in absolute value uniformly in $0 \leq x \leq 1$. (Recall that $h(0) = N - 1$ and $h(x) \sim N$ so the variation in h is about 1 over the interval.)

What remains now is to solve for $F(t)$ and c. From (6) we directly calculate

$$F^{-1}(x) = c[1 + \int_0^x h(y)dy]$$
$$= c[u + 1 + \frac{e^{-u} - 1}{N} + \frac{1 - (u^2 + u + 1)e^{-u}}{N^2} + 0(1/N^3)], \tag{16}$$
$$u = Nx$$

When $x = 0(1)$, (16) reduces to, with exponentially small error

$$F^{-1}(x) = c[Nx + 1 - 1/N + 1/N^2 + 0(1/N^3)], \qquad x = 0(1) \tag{17}$$

We apply $F^{-1}(1) = 1$ to (17) and find that c satisfies

$$\begin{aligned} c &= [N + 1 - 1/N + 1/N^2 + O(1/N^3)]^{-1} \\ &= 1/N - 1/N^2 + 2/N^3 - 4/N^4 + O(1/N^5) \end{aligned} \tag{18}$$

When N = 3, (18) gives $c \simeq 0.247$ as opposed to the exact $c = 0.266$.

We note that if the approximation $h(x) \sim N$ was used, then we would have obtained $c \simeq (N + 1)^{-1} = 1/N - 1/N^2 + O(1/N^3)$, the first two terms in (18). Thus, in the expansion of c, the effect of the boundary layer does not appear until the third term in the expansion.

Finally, we determine an asymptotic approximation to F(t). We invert by considering separately the inner expansion (16) and outer expansion (17) for $F^{-1}(x)$. We begin with (17). Putting $s = F^{-1}(x)$ in (17), we solve for x, replacing c with its expansion from (18). The result is

$$F(s) \sim (1 + 1/N - 1/N^2 + 1/N^3)s - (1/N - 1/N^2 + 1/N^3), \quad s = O(1) \tag{19}$$

We consider (16) next. We set $s = F^{-1}(x)$ and would like to solve for u in terms of s (where u = Nx). The most convenient way to do this is to first put $s = c(1 + w)$, $w = O(1)$, so that

$$c(1 + w) = c\left[u + 1 + \frac{e^{-u} - 1}{N} + \frac{1 - (u^2 + u + 1)e^{-u}}{N^2} + O(1/N^3)\right], \quad w = O(1)$$

and

$$w = u + \frac{e^{-u} - 1}{N} + \frac{1 - (u^2 + u + 1)e^{-u}}{N^2} + O(1/N^3), \qquad w = O(1) \tag{20}$$

gives w in terms of u. An asymptotic expansion for u in terms of w is obtained by assuming an expansion for u(w) in the form $u = u_0(w) + u_1(w)/N + u_2(w)/N^2 + \dots$, plugging into (20), and equating powers of 1/N, up through $1/N^2$. We obtain

$$u = Nx = NF(s) = w + \frac{1 - e^{-w}}{N} + \frac{(w^2 + w + 2)e^{-w} - e^{-2w} - 1}{N^2} + O(1/N^3), \quad w = O(1)$$

and finally

$$F(s) \sim \frac{w}{N} + \frac{1 - e^{-w}}{N^2} + \frac{(w^2 + w + 2)e^{-w} - e^{-2w} - 1}{N^3} \tag{21}$$

where $w = s/c - 1$

The approximation (21) is in fact valid for all $s \in [c,1]$; it is easily checked using (18) that when $s = 0(1)$, (21) reduces to (19). Thus (21), where c is to be replaced via (18), is the final approximation to $F(s)$. A three-term uniform asymptotic approximation to $F'(s)$, the equilibrium probability density, may be obtained by differentiating (21).

This completes the discussion of Example 1.

5. A GENERALIZATION

Without loss of generality, we were able to solve Example 1 by considering the payoff function $t_i - t_{i-1}$ for the player choosing $t = t_i$. We now analyze a generalization of this problem by assuming a payoff function $G(t_i - t_{i-1})$ for the player choosing t_i. The asymptotic analysis of the integral equation which arises in this game is more difficult than in Example 1; it is not possible to reduce the equation to a linear one and we must work directly with the nonlinear equation.

EXAMPLE 2. On the interval $[0,T]$ let N players independently choose a $t \in [0,T]$. If the ordered choices are denoted by t_i, $i = 1,\ldots,N$ then the payoff to the player choosing t_i is $G(t_i - t_{i-1})$ where we assume that G is monotone increasing, $G(0) = 0$, and $G'(t) = g(t)$.

Using arguments similar to those that led to (2), we obtain

$$C(s) = \int_0^s [1 - F(s) + f(s - r)]^{N-1} g(r)dr \tag{22}$$

where F is the assumed equilibrium distribution, $C(s)$ is the payoff to a single player from a choice $t = s$ when his $N - 1$ opponents use F.

We assume the same structure for F as before: $F(c) = 0$, $F'(t) > 0$ for $c \leq t \leq T$. Again, c must be determined. The equilibrium payoff is then $G(c)$. The integral equation is obtained by requiring that $C(s) = C(c) = G(c)$ for $c \leq s \leq T$. For convenience we define $s = x + c$, $H(x) = F(x + c)$, $x \in [0, T - c]$. The integral equation is then

$$G(c) = [1 - H(x)]^{N-1} [G(c + x) - G(x)] + \int_0^x [1 - H(x) + H(x - r)]^{N-1} g(r)dr \tag{23a}$$

with additional requirements on H:

$$H(0) = 0, \qquad H(T - c) = 1 \tag{23b}$$

There does not seem to be any convenient transformation simplifying (23a); nor is the solution for even, say, $N = 2$ readily obtainable. We will proceed then directly to the case of large N; for small N only a numerical analysis seems possible.

For this game we will seek the asymptotic behavior of $F(t)$ as N goes to infinity and scale T with N so that $T = N$. This scaling allows us to treat situations in which N is large but the value of T is such that $g(t)$ varies by $O(1)$ relatively between time choices of the players. (In other words the asymptotic behavior of F will not just depend on the local nature of $g(t)$ near $t = 0$, as would be the case if we left T fixed and let $N \to \infty$.) We assume then that $T = N$ and $N \to \infty$. The case of T fixed can be handled with similar methods. The analysis carried out below will require that a growth condition be put on $g(t)$: $g(t)e^{-pt} \to 0$ as $t \to \infty$ for all $p > 0$. Thus $g(t)$ is enjoined from growing exponentially fast.

We will develop the first order solution below and indicate how further terms could be obtained.

Outer solution: This is the region $H(x) = O(1)$, $x = O(N)$. We assume that H is approximately uniform in this region which suggests $H'(x) \sim 1/N$. If we put $H'(x) = 1/N$ into (23a), the right hand side (with exponentially small error) equals

$$\int_0^x (1 - r/N)^{N-1} g(r)dr = \int_0^\infty e^{-r} g(r)dr + O(1/N)$$

We see that (23a) is satisfied to lowest order if c is picked so that

$$G(c) = \int_0^\infty e^{-r} g(r)dr \tag{24}$$

Inner solution: This is the region $H(x) = O(1/N)$, $x = O(1)$. We put $H(x) = H^*(x)/N$ to reflect the scaling of H in this region. Then in (23a) we use the first order approximations $[1 - H^*(x)/N]^{N-1} \sim \exp(-H^*(x))$; $[1 - H^*(x)/N + H^*(x - r)/N]^{N-1} \sim \exp(-H^*(x) + H^*(x - r))$. Then (23a) becomes, after rearrangement:

$$e^{H^*(x)} = \frac{G(x + c) - G(x)}{G(c)} + \int_0^x e^{H^*(x - r)} \frac{g(r)}{G(c)} dr \tag{25}$$

We observe that (25) is a linear convolution type integral equation for e^{H^*} (also called a renewal equation). Obviously, it cannot be solved without knowing $g(r)$, and even given $g(r)$, a numerical solution may be necessary.

The asymptotic behavior of $H^*(x)$ as $x \to \infty$ can be obtained using well known limit theorems for renewal type integral equations such as (25). (see for instance [1].) The appropriate result for (25) is:

As $x \to \infty$, $H^*(x) - x \to \ln(b)$ where the constant b is given by

$$b = \frac{\int_0^\infty [G(x + c) - G(x)]e^{-x}dx}{\int_0^\infty xg(x)e^{-x}dx} \tag{26}$$

Equation (26) is precisely the information needed to match the behavior of the inner solution with that of the outer solution:

$$\text{As } x \to \infty, \qquad H(x) = H^*(x)/N \sim x/N + \ln(b)/N$$

When $x = O(N)$ this matches up with the outer solution, $H'(x) = 1/N$ to first order, and in fact allows us to determine the next order behavior of $H'(x)$. Assuming that $H'(x)$ is constant with exponentially small error in the outer expansion (this assumption can be justified), the requirement that $H(T - c) = H(N - c) = 1$ means that to second order

$$H(x) \sim x/N + \ln(b)/N + x(c - \ln(b))/N^2$$

We may take, finally, as our uniformly valid asymptotic solution

$$H(x) \sim H^*(x)/N + x(c - \ln(b))/N^2$$

where $H^*(x)$ is the solution of (25). This approximation should give $H(x)$ correct within an error $O(1/N^2)$ uniformly in the interval $[0,T]$. The approximation

$$H'(x) \sim [H^*(x)]'/N$$

gives a uniformly valid approximation to $H'(x)$ having error $O(1/N^2)$.

This completes the first order solution. We make some remarks on the solution to second order: the information obtained from matching, that $H'(x) \sim 1/N + (c - \ln(b))/N^2$ is substituted into (23a) and a second order requirement on $G(c)$ is obtained, furnishing the equilibrium payoff to second order. One must then seek a second term in the inner solution and perform the next order matching. The process continues with the application of $H(T - c) = 1$, giving a third order term in the expansion of $H'(x)$; and further terms in the expansions of $G(c)$ and the inner solution are obtained by the previous route. The procedure is bound to become rather messy as further terms are sought, but the required steps in the analysis are fairly clear. We note the interesting cyclical manner in which matching inner with outer and employing $H(T - c) = 1$ allows higher order terms to be obtained.

We complete the discussion with a simple example: $G(t) = t^2$. From (24) we obtain $G(c) = c^2 = 2$ to first order. The integral equation for H^*, (25), has solution

$$H^*(x) = \ln\left[\frac{1 - \sqrt{2}}{2} e^{-x} + \frac{1 + \sqrt{2}}{2} e^{x}\right] \qquad (27)$$

We see directly from the expression for H^* that as $x \to \infty$, $H^* \sim x + \ln[(1 + \sqrt{2})/2]$ so that $b = (1 + \sqrt{2})/2$. This result can also be obtained of course from (26).

To first order, the equilibrium probability density, $H'(x)$, may be obtained by differentiating (27):

$$NH'(x) \sim (d/dx)H^*(x) \sim 1 + \frac{2(\sqrt{2} - 1)}{(1 + \sqrt{2})e^{2x} + (2 - \sqrt{2})} \tag{28}$$

We observe that $NH'(0) \sim \sqrt{2}$ while $NH'(x) \to 1$ as $x \to \infty$. Thus H' undergoes a relative change of $O(1)$ in an interval of size $O(1/N)T$. In the linear case of $G(t) = t$, the relative change in the function $h(x)$ was $O(1/N)$ over the boundary layer. The effect of the boundary layer in the nonlinear case is thus, generally speaking, more pronounced, in fact, an order of magnitude larger.

6. CONCLUSIONS

Using techniques from singular perturbation theory, we have obtained the asymptotic behavior of the symmetric equilibrium in two N-person games. The work we have done represents a somewhat novel application of these advanced methods of analytic approximation to problems in game theory; and furnishes one of the few examples of integral equations which have been treated via singular perturbation methods.

The type of analysis carried out here can no doubt be applied to similar N-person games. The basic requirement is that the players' possible actions be drawn from a continuum or some other continuous region in R^n. The expected payoffs to the players will then involve integrals and lead to integral equations for the equilibrium which then hopefully can be analyzed when N is large. In general of course such problems can become very difficult; the problems we treated were relatively simple in that the pure strategies were drawn from a one-deminsional continuum and the probability distribution of the expected payoff was easily calculated.

REFERENCE

1. S. Karlin and H. M. Taylor, *A First Course in Stochastic Processes*, 2nd. Ed., Academic Press, New York 1975, Chapter 5, Theorem 5.1.

SUMMARY ANALYSIS OF AN AGE-DEPENDENT, NONLINEAR MODEL OF SEED DISPERSAL

Richard H. Elderkin

Department of Mathematics
Pomona College
Claremont, California

There has long been a healthy interest in mathematical models of populations and epidemics. Many of these, including the early models of Malthus and Verhulst, are in the mathematical form of an ordinary differential equation

$$\frac{dx}{dt}(t) = f(t,x(t)) \tag{1}$$

where x may be a vector state. Because of the possibility of increased sophistication in modelling, functional differential equations (cf. [1,4]) of the form

$$\frac{dx}{dt}(t) = f(t,x_t) \tag{2}$$

have seen dramatic increase of use. Whereas in (1) the derivative dx/dt at time t depends only on t and the present state x(t), in (2) it is also allowed to depend on the past history x_t of x. When x is a measurement on some population, whether its size or its state of infectivity, etc., it is frequently desirable to allow x to vary as a function of age a as well as time. Since age increments with time, the left hand sides of (1) and (2) are replaced by

$$Dx = \frac{\partial x}{\partial a} + \frac{\partial x}{\partial t}$$

If a model takes the form

$$Dx(a,t) = f(a,t,x(a,t)) \tag{3}$$

then it is easy to see that the problem can be reinterpreted as a parameterized family of ordinary differential equations, one for each value of the difference t - a, where (3) is replaced by

$$\frac{dx}{ds}(a + s, t + s) = f(a + s, t + s, x(a + s, t + s)) \tag{4}$$

However, if Dx(a,t) not only has the dependency indicated (3), but also depends on x over all values of a, and perhaps for previous history in t, then the technique of (4) is no longer appropriate. The recent models in [2, 3, 5] fall into this category. In the particular use of [2], a model for malaria proposed by K. Dietz was given as

$$\begin{aligned} Du &= -uv + \int_0^\infty k(a)w(a,t)da \\ Dv &= \alpha u + \beta(v - v_o) \\ Dw &= \gamma ue^{-Tv} - \delta w \end{aligned} \tag{5}$$

where all parameters and functions are to have nonnegative values, and the state (u,v,w) is specified when $t = 0$ and $a \geq 0$, or $a = 0$ and $t \geq 0$. Our interest here is not so much in the derivation of (5) as in its mathematical form. In [2] the analysis of (5) was limited to its steady state (time independent) form, without any consideration of the stability of the steady state.

It is our purpose in this article to consider a model for seed dispersal, proposed by S. Levin, which is of the same mathematical form as (5) but simpler. Hence our goals are twofold: first to understand the model for its own sake, and second to develop tools which may be applicable to further analysis of models like (5). Our model is

$$Du(a,t) = f(u(a,t)) - \delta u(a,t) + \alpha\delta \int_0^1 u(a,t)da \tag{6}$$

subject to specification of $u(a,t)$ when $t = 0$ and $a \geq 0$ or $a = 0$ and $t \geq 0$. In brief, the right hand side of (6) can be loosely justified as follows. The population u is measured in seed equivalents at each time t across an environment which is composed of a mosaic of patches, parameterized by their ages, a. Each patch goes through a cycle of time length one, immediately after which the population of the patch crashes to zero and a new cycle begins. (Hence $u(0,t) \equiv 0$ in the case of strictly biological interest.) The patch ages are out of phase, with all ages equally represented. In the right hand side of (6), $f(u)$ represents the local growth law in the absence of effects of dispersal ($f(0) = 0$), δ is the proportion of a local population which is given up to disperse, and α is the probability of successful establishment in a new patch for dispersed population.

We rewrite the model in the form

$$Du(a,t) = F(u(a,t)) + k \int_0^1 u(a,t)da \tag{7a}$$

$$u(a,0) = u^o(a,0), \qquad 0 < a \leq 1 \tag{7b}$$

$$u(0,t) = u^o(0,t), \qquad t \geq 0 \tag{7c}$$

For the duration of this article, we assume that u^o is nonnegative, measurable and essentially bounded, that $k > 0$, that $F(0) = 0$, and that F is continuously differentiable, with $F'(u)$ nonincreasing and $F'(u) < -k$ for u suitably large.

That which follows summarizes material which will appear elsewhere. We will try to indicate the central themes of propositions and proofs, but arguments are intuitive only and technically incomplete.

THEOREM 1. The problem (7) has a unique solution, defined for $0 \leq a \leq 1$ and $t \geq 0$, which is measurable and bounded almost everywhere.

Proof. The proof is based on the same Picard iteration or contraction mapping principle as is used frequently in the finite dimensional case.

THEOREM 2. The integral $\int_0^1 u(a,t)da$ is continuous in t so that $u(a + s, t + s)$ is smooth in s (C^{r+1} if F is C^r). Moreover $u(a,t)$ is as smooth (jointly in a and t) as $u^o(a,0)$ for $t < a$ and $u^o(0,t)$ for $t \geq a$. A necessary and sufficient condition that u be continuous along $a = t$ is $u^o(0^+,0) = u^o(0,0^+)$. A necessary and sufficient condition that u be C^1 along $a = t$ is

$$\frac{d}{dt} u^0(0,t)\Big|_{t = 0^+} + \frac{d}{da} u^o(a,0)\Big|_{a = 0^+} = F(u^o) + k \int_0^1 u^o(a,0)da$$

where $u^o = u^o(0^+,0) = u^o(0,0^+)$.

Proof. In the proof of Theorem 1, it is shown that the solution of (7) can also be obtained for small negative times, say for $-T < t \leq 0$. One then observes that smoothness of $u^o(a,0)$ for $0 < a < T$ is equivalent to smoothness of $u(0,t)$ for $-T < t < 0$. Furthermore $u(a,t)$ is as smooth near (a_o,t_o) as $u(0,t)$ is smooth near $t = t_o - a_o$. Hence smoothness of $u(a,t)$ along $t = a$ is determined by the smoothness of $u(0,t)$ for $t \geq 0$ and $u^o(a,0)$ for $0 < a < T$. The above calculations result, as do similar ones for higher degrees of differentiability.

THEOREM 3. Let $u_i(a,t)$ solve (7a) with $u_i(a,0) = u_i^o(a,0)$, $u_i(0,t) = u_i^o$ (constant), $i = 1, 2$. If $u_1^o \leq u_2^o$ and $u_1(a,0) \leq u_2(a,0)$ for $0 \leq a \leq 1$, then $u_1(a,t) \leq u_2(a,t)$ for $0 \leq a \leq 1$ and $t \geq 0$.

Proof. The crux of the argument can be seen by supposing that $u_1(a,t) < u_2(a,t)$ for $0 \leq a \leq 1$ and $0 \leq t < T$ with $u_1(a_o,T) = u_2(a_o,T)$ for some $a_o > 0$. Suppose also that $\int_0^1 u_1(a,T)da < \int_0^1 u_2(a,T)da$. Then

$$D(u_2 - u_1)(a_o,T) = k \int_0^1 (u_2 - u_1)(a,T)da > 0$$

which necessitates $u_1(a_o - s, T - s) > u_2(a_o - s, T - s)$ for small $s > 0$, a contradiction.

Having established basic properties of time varying solutions $u(a,t)$, we ask if there are time independent ones, or steady state solutions, which might serve as a limit for $u(a,t)$ as $t \to \infty$ if $u^o(0,t)$ is constant. Such a solution $u(a)$ must satisfy the conditions

$$\frac{du}{da} = F(u) + k \int_0^1 u(a)da \tag{8a}$$

$$u(0) = u^o \tag{8b}$$

where $u^o \geq 0$ in general, and $u^o = 0$ is of biological interest. Notice that (8a) is really a functional differential equation, relating "present" rates of change to "future" states. It is perhaps better viewed as a requirement of consistency between the local and global states, emphasized by rewriting it in the form

$$\frac{du}{da}(a) + F(u(a)) = k \int_0^1 u(a)da$$

To consider the existence and uniqueness of solutions for (8), it is helpful to consider the parameterized family of initial value problems

$$\frac{du}{da} = F(u) + kc \tag{9a}$$

$$u(0) = u^o \tag{9b}$$

letting the unique solution be denoted by $u(a;c)$. Since we only use this notation temporarily while examining the existence and uniqueness of solutions to (8), there should be no confusion with $u(a,t)$ as a solution of (7). We define

$$G(c) = \int_0^1 u(a;c)da, \qquad c \geq 0$$

so that $u(a;c)$ solves (8) if and only if $G(c) = c$. Now G is continuous and $G(0) \geq 0$. By setting the right hand side of (9a) to zero and solving for c, one is able to show that eventually $G(c) < c$. One shows that $G'(c)$ is monotonically decreasing by using the variational equation for $\partial u/\partial c$ derived from (9a). When $u^o > 0$, it is easily seen that $G(0) > 0$ and so we have the following theorem.

THEOREM 4. If $u^o > 0$ the steady state problem (8) has a unique solution.

In the case $u^o = 0$, there is always the trivial solution $u \equiv 0$, corresponding to $G(0) = 0$. Clearly there will be a (necessarily unique) nontrivial steady state if and only if $G'(0) > 1$. Again one analyzes the variational equation derived from (9). When $u^o = 0$, it is simple enough to make explicit calculations.

THEOREM 5. If $u^o = 0$, the steady state problem (8) has at most one nontrivial solution. Such a solution exists if and only if

$$F'(0)^{-2}[e^{F'(0)} - (1 + F'(0))] > 1/k$$

If $u(a,t)$ solves (7) with $u^o(0,t) \equiv u^o$ and $\bar{u}(a)$ solves the corresponding steady state problem (8), we naturally inquire if $u(a,t)$ converges to $\bar{u}(a)$ as $t \to \infty$. Considering that the right hand side of (7a) is linear in the function state $u(\cdot,t)$ and stable quadratic (in the simplest cases) in the local state $u(a,t)$, we have reasonable basis for hope. Henceforth we consider (7c) only in the form $u^o(0,t) \equiv u^o$ (constant) for $t \geq 0$.

Since the proofs of the stability theorems are rather similar in form, we will state all three theorems and then give a sketch of the argument in a simple, but illustrative, special case.

THEOREM 6. If $F'(0) + k < 0$ and $u^o = 0$ then the steady state $\bar{u}(a) \equiv 0$ is globally asymptotically stable, that is, given any nonnegative essentially bounded initial data $u^o(a,0)$, $u(a,t) \to 0$ uniformly for $a \in [0,1]$ as $t \to \infty$.

THEOREM 7. Let $u^o = 0$ and let $\bar{u}(a)$ be a nontrivial solution of (8) ($F'(0) + k > 0$). If $\int_0^1 u^o(a,0)da > 0$ then the resulting solution $u(a,t)$ of (7) converges to $\bar{u}(a)$ as $t \to \infty$, uniformly for $a \in [0,1]$.

THEOREM 8. If $u^o > 0$ then every solution $u(a,t)$ of (7) converges to the steady state uniformly for $a \in [0,1]$.

We should interject a remark about the role of the quantity $F'(0) + k$ which appears in Theorems 6 and 7. Notice that it is rather like the linearization of the right side of (7a) for $u \simeq 0$. So we expect the trivial steady state to be stable when $F'(0) + k > 0$. Furthermore, we can compute that $F'(0) + k > 0$ is a necessary condition for the existence of a nontrivial steady state when $u^o = 0$. Indeed, when $F'(0) < 0$ we have from Theorem 5

$$-F'(0)^{-1} > F'(0)^{-2}[e^{F'(0)} - (1 + F'(0))] > k^{-1}$$

Theorem 3 on the monotonicity is the keystone for our stability arguments. A typical application will run roughly as follows. Let $\bar{u}(a)$ be some steady state solution, $\bar{u}(0) = u^o$, and let initial data $u^o(a,0)$ generate the solution $u(a,t)$ to (2) and (3). It will be convenient to consider alternate initial data, $u^o_+(a,0)$ and $u^o_-(a,0)$, of some extremely simple form and such that $u^o_-(a,0) \leq \min\{u^o(a,0), \bar{u}(a)\}$, $u^o_+(a,0) \geq \max\{u^o(a,0), \bar{u}(a)\}$. Letting $u^o_\pm(a,0)$ generate respective solutions $u_\pm(a,t)$, we have $u_-(a,t) \leq u(a,t) \leq u_+(a,t)$ from the monotonicity theorem. The simple forms of $u^o_\pm$ make it possible to show convergence of $u_\pm$ to $\bar{u}$. Because of the way u is sandwiched between u_+ and u_-, it also must converge to $\bar{u}$.

For an illustrative special case, let us assume that $u^o > 0$ and that $0 \leq u^o(a,0) \leq c$, $0 \leq a \leq 1$. As we mentioned above, there is no loss of generality in separately considering the cases $u^o(a,0) \equiv 0$, $u^o(a,0) \equiv c$. Let's consider $u^o(a,0) \equiv c$. We may as well assume that $F(c) + k < 0$, since this is the case for all sufficiently large c. Now u is smooth except across $a = t$, and if we can show that $\partial u/\partial t(a,t) < 0$ for $0 < a \leq 1$, $t \geq 0$, $a \neq t$, then fairly standard arguments show that $u(a,t)$ must decrease monotonically, and hence uniformly, to the steady state. Certainly things begin well at $t = 0$, for

$$\frac{\partial u}{\partial t}(a,0) = c(F(c) + k) < 0$$

Moreover

$$\frac{d}{dt} \int_0^1 u(a,t)da \Big|_{t=0} = \int_0^1 \frac{\partial u}{\partial t}(a,0)da + u^o - u(0^+,0) < 0$$

At this point one must examine the relation between $\partial u/\partial t(a,t)$ and $d/dt \int_0^1 u(a,t)da$ for $t > 0$. Observing that $u(a + s, t + s)$ satisfies the ordinary differential equation

$$\frac{du}{ds} = F(u) + k \int_0^1 u(a,t + s)da$$

one views t as a parameter and uses the variational equation from this o.d.e. to obtain the desired realtion between $\partial u/\partial t$ and $d/dt \int_0^1 u(a,t)da$. We find that under the above conditions neither $\partial u/\partial t(a,t)$ for $0 < a \leq 1$, $t \geq 0$, $t \neq a$ nor $d/dt \int_0^1 u(a,t)da$ can be nonnegative. This completes the argument.

ACKNOWLEDGMENT

This work was supported by National Science Foundation Grant MCS7903497.

REFERENCES

1. R. Bellman and K. Cooke, *Differential Difference Equations*, Academic Press, New York 1952.

2. R. Elderkin, D. Berkowitz, F. Farris, C. Gunn, F. Hickernell, S. Kass, F. Mansfield, and R. Taranto, "On the steady state of an age dependent model for malaria," *Nonlinear Systems and Applications*, Academic Press, New York 1977, 491-521.

3. M. Gurtin and R. MacCamy, "Non-linear age-dependent population dynamics," *Arch. Rat. Mec. Anal.* 54 (1974), 281-300.

4. J. Hale, *Theory of Functional Differential Equations*, Springer-Verlag, New York 1977.

5. K. Swick, "A Model of single species population growth," *SIAM J. Math. Anal.*, 7 (1976) 565-576.

INTEGRODIFFERENTIAL EQUATIONS IN BANACH SPACE

William E. Fitzgibbon*

Department of Mathematics
University of Houston
Houston, Texas

1. INTRODUCTION

We are concerned with the application of the theory of semigroups of linear operators and the theory of abstract quasilinear parabolic equations to partial integrodifferential equations. We consider equations of parabolic and hyperbolic type. Specifically we consider an abstract parabolic equation of the form:

$$\dot{x}(t) + A(x(t))x(t) = \int_0^t g(t - s)K(x(s), x(s))ds + f(t) \tag{1.1}$$

Here $A(\cdot)$ is an unbounded quasilinear operator mapping a Banach space X crossed with itself to itself; for fixed $x \in X$, $A(x)$ is the infinitesimal generator of an analytic semigroup on X. The function $K : X \times X \to X$ is an unbounded nonlinear operator; g is a scalar valued function and $f : R \to X$.

We also consider an abstract hyperbolic equation with infinite memory:

$$\begin{aligned} \ddot{x}(t) + Ax(t) &= \int_{-\infty}^t g(t - s)K(x(s), x(s))ds + f(t) \\ x(\theta) &= \phi(\theta) \qquad \theta \in (-\infty,0) \end{aligned} \tag{1.2}$$

**Current affiliation*: Department of Mathematics, University of California, San Diego, La Jolla, California.

Here A is a positive self adjoint operator on a Hilbert space H and $K : H \times H \to H$ is a nonlinear operator.

Throughout this paper X shall denote a Banach space; if A is a closed densely defined linear operator on X, X_A will denote the domain of A endowed with the graph norm

$$\| x \|_A = \| x \| + \| Ax \|$$

Similarly C_A will denote the space of bounded uniformly continuous functions mapping the interval $(-\infty, 0)$ to X_A which are given the supremum norm

$$\| \phi \|_{C_A} = \sup_{\theta \in (-\infty,0)} \| \phi(\theta) \|_A$$

2. PARABOLIC CASE

We begin with a local existence theorem. We place the following conditions on $A(\cdot) : X \times X \to X$.

(A.1) The operator $A_0 = A(x_0)$ is a closed densely defined linear operator in X and

$$\| (\lambda I - A_0)^{-1} \| \leq C(1 + |\lambda|)^{-1} \text{ for all } \lambda \text{ with } \mathrm{Re}\lambda \leq 0$$

(A.2) For some $\alpha \in (0,1)$ and $R > 0$ and for any $v, w \in X_{A_0^\alpha}$ with $\| v \| < R$ and $\| w \| < R$ the operators $A(v)$ and $A(w)$ are defined. Furthermore:

$$\| \{A(v) - A(w)\} A^{-1}(w) \| \leq C(R) \| v - w \|_{A_0^\alpha}$$

Conditions (A.1) and (A.2) are often called the Tanabe-Sobolevskii conditions; they are developed at length in Friedman [12] or Sobolevskii [20]. We require that $K : X \times X \to X$ satisfy the following conditions:

(N) There exists a function $h : X_A \to R$ uniformly continuous on bounded subsets of X_A so that

$$\| K(x_1,y) - K(x_2,y) \| \leq h(y) \| x_1 - x_2 \|_{A_0^\alpha}$$

$$\text{for } x_1,\ x_2 \in X_{A_0^\alpha}, \quad y \in X_{A_0}$$

and

$$\| K(x,y_1) - K(x,y_2) \| \leq h(x) \| y_1 - y_2 \|_{A_0}$$

$$\text{for } x \in X_{A_0},\ y_1,\ y_2 \in X_{A_0^\alpha}$$

We have the following local existence theorem.

THEOREM 2.1. Let $A(\cdot)$ and $K(\cdot,\cdot)$ satisfy conditions (A.1) and (A.2) and N respectively. If $g : R \to R$ is continuously differentiable with Hölder continuous derivative and $f : R \to X$ is continuously differentiable with Hölder continuous derivative then there exists $t_0 = t(x_0)$ and a unique function $x(\cdot) : [0,t_0] \to X_{A_0}$ which satisfies

$$\begin{aligned} &\dot{x}(t) + A(x(t))x(t) = \int_0^t g(t - s)K(x(s), x(s))ds + f(t) \\ &\qquad \text{for } t \in [0,t_0] \qquad\qquad (2.2) \\ &x(0) = x_0 \end{aligned}$$

Indication of proof. The proof consists of a double application of the contraction mapping theorem. We first apply it to obtain a class of approximating functions and then apply it again to obtain our desired solution.

We let $t^* > 0$ and let S be the class of Lipschitz continuous functions from $[0,t^*]$ to X such that

$$\| v(t) - v(\tau) \| \leq k|t - \tau|$$

$$v(0) = A_0^\alpha x_0$$

We define

$$A_v(t) = A(A_0^{-\alpha}v(t))$$

and argue that $\{A_v(t)|0 \leq t \leq t^*\}$ generates a linear evolution system $\{W_v(t,s)|0 \leq s \leq t \leq t^*\}$ on X. We consider $x(\cdot) \in C([0,t^*],X_{A_0})$ and define a mapping G by the equation

$$(Gx)(t) = W_v(t,0)x_0 + \int_0^t W_v(t,s) \int_0^s g(s-r)K(x(r), x(r))dr\, ds + \int_0^t W_v(t,s)f(s)ds \qquad t \in [0,t^*]$$

We utilize the regularity properties of $W_v(t,s)$ to argue that a correct choice of t^* and K guarantees that G is a contraction on a closed subset of the Banach space $C([0,t^*],X_A)$. Thus we are assured the existence of a fixed point $x_v(\cdot)$ satisfying:

$$x_v(t) = W_v(t,0)x_0 + \int_0^t W_v(t,s) \int_0^s g(s-r)K(x(r), x(r))dr\, ds + \int_0^t W_v(t,s)f(s)ds$$

This equation may be differentiated to produce,

$$\dot{x}_v(t) + A_v(t)x_v(t) = \int_0^t g(t-s)K(x_v(s), x_v(s))ds + f(t)$$

We argue that $x_v(t)$ may be bounded in X_A independent of a particular choice of $v(\cdot)$.

We now specify $Q(t,k,\alpha)$ to be the class of Lipschitz continuous functions mapping $[0,t^*]$ to X such that

$$y(0) = A_0^\alpha x_0$$

and

$$\| y(t) - y(\tau) \| \leq K|t - \tau|$$

For $t \in [0,t^*]$ we define

$$w_v(t) = A_0^\alpha x_v(t)$$

and once again use the regularity of $W_v(t,s)$ to show that an appropriate choice of K and t* insures that the mapping H defined by $H_v = W_v$ is a contraction of $Q(t^*,K,\alpha)$ into itself. We let z be the fixed point of the mapping H and argue that $x(t) = A_0^{-\alpha}z(t)$ is the solution of (2.2).

The problem with our approach is that it produces solutions which are local in time. This is due to the fact that $A_x(t) = A(x(t))$ depends on the solution; therefore we cannot use the standard Gronwall approach to produce bounded solutions. However we obtain maximal solution intervals and show that if $t_{max} < \infty$ then

$$\overline{\lim_{t\to t_{max}}} \| (A(x(t)))^{\alpha}x(t) \| = \infty$$

If we modify our equation and require that the unbounded operator outside of the integral be linear we can obtain global existence. Specifically we study the equation:

$$\dot{x}(t) + Ax(t) = \int_0^t g(t - s)K(x(s), x(s))ds + f(t) \tag{2.3}$$

$$x(0) = x_0$$

We place the following requirements on K.

(N.1) There exists a function $h : X_A \to R$ which is uniformly continuous on bounded subsets of X_A such that

$$\| K(x_1,y) - K(x_2,y) \| \le h(y)\| x_1 - x_2 \|^{\alpha}_{A^{\alpha}}$$

for some $\alpha \in (0,1)$, x_1, $x_2 \in X_{A^{\alpha}}$ and $y \in X_A$

and

$$\| K(x,y_1) - K(x,y_2) \| \le h(x)\| y_1 - y_2\|_A$$

for $x \in X_{A^{\alpha}}$, y_1, $y_2 \in X_A$

(N.2) There exists a constant $M > 0$ so that

$$\| K(x,y) \| \leq M(\| x \|^{\alpha}_{A^{\alpha}} + \| y \|_A) \text{ for some } \alpha \in (0,1),\ x \in X_{A^{\alpha}},\ y \in X_A$$

We have the following result:

THEOREM 2.4. Let A satisfy condition (A.1) and assume that A^{-1} is compact. If $K(\cdot,\cdot)$ satisfies (N.1) and (N.2) and $f(\cdot)$ and $g(\cdot)$ satisfy the conditions of Theorem 2.1 then for each $x_0 \in D(A)$ there exists a function $x(\cdot) : [0,\infty) \to X_A$ which satisfies (2.3).

The proof of the foregoing result is very similar to the proof of Theorem (2.1); it consists of two applications of fixed point theorems; however, the Schauder fixed point theorem is applied at the second stage. The global continuation is produced by standard Gronwall arguments. The result appears in detail in [9]. We remark that the fact that $K(\cdot,\cdot)$ is only required to be Hölder continuous in the first place allows the existence of nonunique solutions.

3. A HYPERBOLIC EQUATION

In this section we consider a hyperbolic equation with infinite delay. We require that $K(\cdot,\cdot)$ be Lipschitz continuous with respect to each place when considered as a mapping from $X_A \times X_A$ to X and that it be linearly bounded in the second place. Specifically:

(3.1) There exists a function $h : X_A \to R^+$ which is uniformly continuous on bounded subsets of X_A so that

$$\| K(x_1,y) - K(x_2,y) \| \leq h(y) \| x_1 - x_2 \|_A \text{ for } x_1,\ x_2,\ y \in D(A)$$

$$\| K(x,y_1) - K(x,y_2) \| \leq h(x) \| y_1 - y_2 \| \text{ for } x,\ y_1,\ y_2 \in D(A)$$

and

(3.2) There exists a constant $M > 0$ so that

$$\| K(x,y) \| \leq M(\| y \|_A + 1) \text{ for } x,\ y \in D(A)$$

We place the following assumptions on $f(\cdot) : [0,T] \to X$ and $g(\cdot) : (-\infty,T) \to R$

(3.3) $f(\cdot) : [0,T] \to X$ is continuously differentiable

(3.4) $g : (-\infty,T] \to R$ is continuously differentiable and uniformly bounded on intervals of the form $(-\infty,T)$: g' is Lipschitz continuous and uniformly bounded on $(-\infty,T)$ and $g, g' \in L^1(-\infty,T)$.

We have the following existence theorem:

THEOREM 3.5. Let $-A$ be the infinitesimal generator of a strongly continuous semigroup on a Banach space X. Assume that $K(\cdot,\cdot) : X_A \times X_A \to X$, $f(\cdot) : [0,T] \to X$ and $g : (-\infty,T] \to R$ satisfy (3.2) through (3.4) respectively. If $\phi \in C_A$ then there exists a unique function $x(\cdot) : (-\infty,T] \to X_A$ satisfying

$$\dot{x}(t) + Ax(t) = \int_{-\infty}^{t} g(t - s)K(x(s), x(s))ds + f(t) \text{ for } t \in [0,T]$$

$$x(\theta) = \phi(\theta) \text{ for } \theta \in [-\infty,0], \quad \phi \in C_A$$

The proof of the foregoing theorem consists of a concatenation of techniques for abstract equations with memory appearing in [25] and abstract equations with infinite dalay appearing in [8]. Existence is first established locally via the contraction mapping theorem; classical methods allow the extension of solutions and uniques follows from the Lipschitz properties of the nonlinearities. A detailed proof will appear in [7].

We now restrict our attention to Hilbert spaces. Notation introduced previously carries over with the understanding that we are now working in Hilbert spaces. If A is a positive self adjoint operator on a Hilbert space then $-A$ is known to be the infinitesimal generator of a semigroup on H. We can use the spectral calculus to compute the positive square root $A^{\frac{1}{2}}$ of A. It is well known that $-A^{\frac{1}{2}}$ is also the infinitesimal generator of a semigroup on H and we designate the spaces $H_{A^{\frac{1}{2}}}$ and $C_{A^{\frac{1}{2}}}$ respectively.

We have the following theorem which guarantees existence of solutions to (1.2).

THEOREM 3.6. Let A be a positive self adjoint operator on a Hilbert space H and let $f(\cdot) : [0,T] \to H$ and $g(\cdot)$ satisfy (3.3) and (3.4). Further assume that $K(\cdot,\cdot) : H_A \times H_A \to H$ satisfies (3.1) and (3.2). If $\phi \in C_A$ is such that $\phi \in C_{A^{\frac{1}{2}}}$ then there exists a unique function $x(\cdot) : (-\infty,T] \to H_A$ which satisfies

$$\ddot{x}(t) + Ax(t) = \int_{-\infty}^{t} g(t - s)K(x(s), x(s))ds + f(t)$$

$$x(\theta) = \phi(\theta) \qquad \theta \in (-\infty, 0)$$

Indication of proof. We interpret (3.7) as a first order system and define the Banach space $\hat{X} = H_{A^{1/2}} \times H$ with norm $\| \cdot \|_{\hat{X}}$

$$\left\| \begin{matrix} x_1 \\ x_2 \end{matrix} \right\| = \| x_1 \|_{A^{1/2}} + \| x_2 \|$$

It is well known that the operator $\hat{A}$ specified by the action of the operator matrix

$$\hat{A} = \begin{pmatrix} 0 & 1 \\ A & 0 \end{pmatrix}$$

on $\hat{X}$ is the infinitesimal generator of a strongly continuous group on $\hat{X}$. We define $\hat{K}(\cdot,\cdot) : \hat{X} \times \hat{X} \to \hat{X}$ by

$$\hat{K}(\hat{x},\hat{y}) = \begin{pmatrix} 0 \\ K(x_1,y_1) \end{pmatrix} \text{ where } \hat{x} = \begin{pmatrix} x_1 \\ x_2 \end{pmatrix} \text{ and } \hat{y} = \begin{pmatrix} y_1 \\ y_2 \end{pmatrix}$$

We let

$$\hat{f}(t) = \begin{pmatrix} 0 \\ f(t) \end{pmatrix}$$

and define

$$\hat{\phi}(\theta) = \begin{pmatrix} \phi(\theta) \\ \dot{\phi}(\theta) \end{pmatrix} \qquad \theta \in (-\infty, 0)$$

Theorem 3.5 guarantees the existence of solutions to

$$\dot{\hat{x}}(t) + \hat{A}\hat{x}(t) = \int_{-\infty}^{t} g(t - s)\hat{K}(\hat{x}(s), \hat{x}(s))ds + \hat{f}(t)$$

$$\hat{x}(\theta) = \hat{\phi}(\theta) \qquad \theta \in (-\infty, 0)$$

We obtain a solution to our desired equation by reading off the second coordinate of the above equation.

4. EXAMPLES

We begin with an example of a nonlinear parabolic equation with memory:

$$u_t(x,t) = \sigma(u_x(x,t))_x + \int_0^t g(t - s)\sigma(u_x(x,s))_x ds + h(x,t) \tag{4.1}$$

$$u(0,t) = u(1,t) = 0$$

$$u(x,0) = \phi(x)$$

We require that σ' be uniformly Lipschitz and that there exist constants $m, M > 0$ so that $m \le \sigma'(x) \le M$. We set $X = L^2(0,1)$ and we formally define an operator $A : X \times X \to X$

$$(A(v)u)(x) = -\sigma'(v'(x))u''(x)$$

defined for all functions $u \in W_2^{(2)}(0,1)$ which vanish on the boundary. In [20] it is shown that A satisfies (A.1) and (A.2) for any $\alpha \in (\frac{1}{2},1)$. We define

$$K(v,u) = -A(v)u$$

and verify that $K(\cdot,\cdot)$ satisfies condition N. We define $f : [0,T] \to X$ by $(f(t))(x) = h(t,x)$. It then becomes clear that equation (2.2) provides a local solution to (4.1) in $L^2(0,1)$.

In the linear case we can examine an equation of the form

$$u_t(x,t) = u_{xx}(x,t) + \int_0^t g(t - s)\sigma(u_x(x,s))_x ds + h(x,t) \tag{4.2}$$

$$u(0,t) = u(1,t) = 0$$

$$u(x,0) = \phi(x)$$

We set $X = L^2(0,1)$ and define

$$(Au)(x) = -u''(x)$$

$$D(A) = \{u \in X | u', u'' \in X;\ u(0) = u(1) = 0\}$$

The term $\sigma'(u_x(x,s))u_{xx}(x,s)$ is used to define $K(\cdot,\cdot)$. Here we can relax the Lipschitz continuity of σ' to Hölder continuity. Equation (2.3) will provide global solutions to (4.2) in $L^2(0,1)$. The details of this example appear in [7].

In the hyperbolic case we examine the following integrodifferential equation with infinite delay

$$u_{tt}(x,t) = u_{xx}(x,t) + \int_{-\infty}^{t} g(t-s)(\sigma(u_x(x,s)))_x ds + h(x,t) \tag{4.3}$$
$$u(0,t) = u(1,t) = 0$$
$$u(x,\theta) = \phi(\theta,x) \qquad \theta \in (-\infty,0)$$

Once again $X = L^2(0,1)$; the operator A satisfies the conditions of the preceding example. The expression $\sigma'(v'(x))u''(x)$ is used to define the nonlinear operator $K(\cdot,\cdot)$. We require that there exists an $M > 0$ so that $0 \leq |\sigma'(x)| \leq M$ and that σ' be Lipschitz. For appropriate initial functions ϕ, equation (3.7) is the $L^2(0,1)$ solution of (4.3).

Our techniques can be modified to consider more general elliptic operators in any number of spatial dimensions with a variety of boundary conditions. We can set equations of types (4.1) and (4.2) in L^p and $C(0,1)$. Equation (4.3) must be treated in a Hilbert space.

REFERENCES

1. V. Barbu, "Nonlinear Volterra equations in Hilbert space," *SIAM J. Math. Anal.* 6 (1975), 728-741.

2. M. Crandall, S. -O. Londen and J. Nohel, "An abstract nonlinear Volterra integrodifferential equation," *J. Math. Anal. Appl.* 64 (1978), 701-735.

3. M. Crandall and J. Nohel, "An abstract functional differential equation and a related nonlinear Volterra equation," *Israel J. Math.* 29 (1978), 313-329.

4. C. Dafermos, "An abstract Volterra equation with applications to linear viscoelasticity," *J. Diff. Equations* 7 (1970), 554-559.

5. C. Dafermos, "Asymptatic stability in viscoelasticity," *Arch. Rational Mech. Anal.* 37 (1970), 297-308.

6. C. Dafermos and J. Nohel, "Energy methods for nonlinear hyperbolic Volterra integrodifferential equations," MRC Technical Report #1876, University of Wisconsin, Madison, Wisconsin, 1978.

7. W. Fitzgibbon, "Abstract hyperbolic integrodifferential equations," *Journal Mathematical Analysis and Applications*, (to appear).

8. W. Fitzgibbon, "Nonlinear Volterra equations with infinite delay," *Monat für Math.* 84 (1977), 297-308.

9. W. Fitzgibbon, "Semilinear integrodifferential equations in Banach space," *J. Nonlinear Analysis: TMA* (to appear).

10. W. Fitzgibbon, "Semilinear functional differential equations in Banach Space," *J. Diff. Equations* 29 (1978), 1-14.

11. W. Fitzgibbon, "Stability for abstract Volterra equations involving finite delay," *J. Math. Anal. Appl.* 60 (1977), 429-434.

12. A. Friedman, *Partial Differential Equations*, Holt, Rinehart and Winston, New York, 1969.

13. J. Goldstein, "Semigroups of Operators and Abstract Cauchy Problems," *Lecture Notes*, Tulane University, New Orleans, Louisiana, 1970.

14. M. Gurtin and A. Pipkin, "A general theory of heat conduction with finite wave speeds," *Arch. Rat. Mech. Anal.* 31 (1968), 113-126.

15. M. Heard, "An abstract semilinear hyperbolic Volterra integrodifferential equation," *Integral and Functional Differential Equations*, Marcel Dekker, New York, 1981 (this volume).

16. R. MacCamy, "An integrodifferential equation with applications to heat flow," *Q. Appl. Math.* 35 (1977), 1-19.

17. R. MacCamy, "A model for one dimensional nonlinear viscoelasticity," *Q. Appl. Math.* 35 (1977), 21-33.

18. J. Nunziato, "On heat conduction in materials with memory," *Q. Appl. Math.* 29 (1971), 187-204.

19. A. Pazy, "Semigroups of Linear Operators and Applications to Partial Differential Equations," *Lecture Notes #10*, University of Maryland, Baltimore, Maryland, 1974.

20. P. Sobolevskii, "Equations of parabolic type in Banach Space," *Amer. Math. Soc. Transl.*, Series 2, 49 (1962), 1-62.

21. C. Travis and G. Webb, "An abstract second order semilinear Volterra integrodifferential equation," *SIAM J. Math. Anal.* (to appear).

22. C. Travis and G. Webb, "Existence and stability for partial functional differential equations," *Trans. Amer. Math. Soc.* 200 (1974), 395-410.

23. C. Travis and G. Webb, "Existence, stability and compactness in the α-norm for partial functional differential equations," *Trans. Amer. Math. Soc.* 240 (1978), 255-260.

24. G. Travis and G. Webb, "Partial differential equations with deviating arguments in the time variable," *J. Math. Anal. Appl.* 56 (1976), 397-408.

25. G. Webb, "An abstract semilinear integrodifferential equation," *Proc. Amer. Math. Soc.* 69 (1978), 255-260.

26. G. Webb, "Abstract Volterra integrodifferential equations and a class of reaction-diffusion equations," *Volterra Equations*, Springer Lecture Notes, No. 737 (1979), 295-303.

27. G. Webb, "Volterra integral equations and nonlinear semigroups," *Nonlinear Analysis, TMA* 1 (1977), 415-427.

AN ABSTRACT SEMILINEAR HYPERBOLIC VOLTERRA INTEGRODIFFERENTIAL EQUATION

Melvin L. Heard

Department of Mathematics
University of Illinois at Chicago Circle
Chicago, Illinois

1. INTRODUCTION

The purpose of this paper is to study the Cauchy problem for the abstract semilinear Volterra integrodifferential equation

$$u'(t) + A(t)u(t) = \int_0^t g(t,s,u(s))ds + f(t), \qquad t \geq 0 \tag{1.1}$$

$$u(0) = u_0 \tag{1.2}$$

For a fixed $t \geq 0$, $A(t)$ is assumed to be the negative generator of a strongly continuous semigroup $\{e^{-sA(t)} : s \geq 0\}$ in a Banach space X. The domain $D(A(t))$ of $A(t)$ is allowed to vary with t. Thus our models for (1.1)-(1.2) are Volterra integrodifferential equations of "hyperbolic type."

Our formulation of (1.1)-(1.2) is a direct attempt to generalize some results of G. F. Webb [10, 11] who studied problems similar to (1.1)-(1.2) for the case when $A = A(t)$ does not depend on t. We are basically interested in the existence, uniqueness, continuation and continuous dependence of solutions of (1.1)-(1.2). The dependence of the generator $A(t)$ on time introduces added complications into our analysis. To overcome difficulties, we treat (1.1)-(1.2) as a perturbation problem for the linear equation

$$u'(t) + A(t)u(t) = f(t), \qquad t \geq 0 \tag{1.3}$$

$$u(0) = u_0 \tag{1.4}$$

For basic results on the Cauchy problem (1.3)-(1.4) we rely on the operator—theoretic methods developed by T. Kato [6-9] and J. Goldstein [4, 5].

Our analysis of (1.1)-(1.2) falls naturally into two parts. Part I deals with the case of variable domains and has as its model, in the case of (1.3)-(1.4), the classical symmetric hyperbolic conservation laws. Part II discusses the case when $D(A(t)) = D$ does not depend on t. The model for (1.1)-(1.2) in this case is a second order hyperbolic integro-differential equations with time varying coefficients. These types of equations arise naturally in the study of viscoelasticity. In this connection, our results offer a generalization of some work of C. M. Dafermos [1].

In this note we only report on results contained in Part I. A complete report will appear elsewhere.

2. EXISTENCE, UNIQUENESS AND CONTINUATION

Let X and Y be Banach spaces and let $\mathcal{L}(Y,X)$ denote the Banach space of all bounded linear operators from Y to X. In what follows, we use the terminology of [8, 9]. Let $\{A(t) : 0 \leq t < \infty\}$ be a family of closed linear operators with domains $D(A(t))$ dense in X. Assume that for each $t > 0$, the operator $-A(t)$ is the infinitesimal generator of a strongly continuous semigroup in X. Assume also that $Y \subset X$ and the embedding $Y \to X$ is dense and continuous. We make the following additional hypotheses on equation (1.1):

(A1) $\{A(t) : 0 \leq t < \infty\}$ is quasi-stable on each bounded interval $[0,T]$ with stability index $\{M,\beta\}$.

(A2) $Y \subset D(A(t))$ for all $t \geq 0$ and $A(t)$ is norm continuous (Y,X).

(A3) For each $T > 0$ there is a family $\{S(t) : 0 \leq t \leq T\}$ of linear isomorphisms from Y to X such that

$$S(t)A(t)S(t)^{-1} = A(t) + B(t) \qquad \text{a.e. on } [0,T]$$

in the strict sense of domain inclusions. The operator $B(t)$ belongs to $\mathcal{L}(X)$, is strongly measurable on $[0,T]$ with $\| B(t) \|_X$ upper-integrable. Also, there is a strongly measurable function $\dot{S} : [0,T] \to \mathcal{L}(Y,X)$ with $\| \dot{S}(t) \|_{Y,X}$ upper-integrable, such that S is an indefinite strong integral of $\dot{S}$.

(A4) f belongs to $C(0,\infty;X) \cap L^1_{loc}(0,\infty;Y)$

(A5) Let $\Delta = \{(t,s) : 0 \leq s \leq t < \infty\}$, then there is a nonempty open subset W of Y such that $g : \Delta \times W \to Y$ is continuous. Furthermore, for each $y \in W$ there is a constant $r > 0$ and a positive continuous function $b(t,s)$ defined on Δ such that if $B_Y(y;r) = \{z \in Y : \| z - y \|_Y \leq r\}$ then $B_Y(y;r) \subset W$ and

$$\| g(t,s,y_1) - g(t,s,y_2) \|_Y \leq b(t,s) \| y_1 - y_2 \|_Y \tag{2.1}$$

for all $y_1, y_2 \in B_Y(y;r)$ and all $(t,s) \in \Delta$.

By Theorem I of [6] there is a unique evolution operator $\{U(t,s) : (t,s) \in \Delta\}$ having the following properties:

(a) U is strongly continuous from Δ to $\mathcal{L}(X)$, with $U(s,s) = I$ for all $s \geq 0$.

(b) $U(t,r)U(r,s) = U(t,s)$ for all $s \leq r \leq t$.

(c) $U(t,s)Y \subset Y$ and U is strongly continuous from Δ to $\mathcal{L}(Y)$.

(d) For each fixed $y \in Y$ we have

$$\frac{\partial}{\partial t} U(t,s)y = -A(t)U(t,s)y$$

$$\frac{\partial}{\partial s} U(t,s)y = U(t,s)A(s)y$$

for all $(t,s) \in \Delta$ and these derivatives are continuous in the X-norm.

Our first result concerns the existence of solutions of (1.1)-(1.2).

THEOREM 1. Let (A1)-(A5) hold and let $u_o \in W$ be given. Then there exists $T = T(u_0) > 0$ and a unique function u belonging to $C(0,T;Y) \cap C^1(0,T;X)$

such that u satisfies (1.1)-(1.2) on [0,T]. Furthermore, if W = Y in hypothesis (A5) and if (2.1) holds for all y_1, $y_2 \in Y$, then the solution u exists on $(0,\infty)$.

The proof of Theorem 1 is accomplished by showing that the operator

$$\Phi v(t) = U(t,0)u_0 + \int_0^t U(t,s)\,\{\int_0^s g(s,\tau,v(\tau))d\tau + f(s)\}ds$$

defined for $v \in C(0,T;B)$, where $B = B_Y(u_0;r)$, has a unique fixed point if $T = T(u_0)$ is sufficiently small.

To discuss noncontinuable solutions of (1.1) we require an additional hypothesis on g.

(A6) g is Lipschitz continuous on bounded subsets of $\Delta \times W$, i.e., if B is a bounded subset of W and Δ_0 is a compact subtriangle of Δ then there is a constant $C > 0$ such that if y_1, $y_2 \in B$ and (t_1,s_1), $(t_2,s_2) \in \Delta_0$ then

$$\| g(t_1,s_1,y_1) - g(t_2,s_2,y_2) \|_Y \le C(\| y_1 - y_2 \|_Y + |t_1 - t_2| + |s_1 - s_2|)$$

THEOREM 2. Assume (A1)-(A6) hold and let u be a solution of (1.1) on a maximal interval of existence [0,d). If $d < +\infty$ then for each closed bounded set $B \subset W$ there is a sequence $\{t_n\}$ such that $t_n \to d^-$ and $u(t_n) \notin B$ for all $n \geq 1$.

COROLLARY. Assume that (A1)-(A6) hold. Assume also that W = Y in (A5) and (A6). Let u(t) be a solution of (1.1) on a maximal interval of existence [0,d). If $d < +\infty$ then $\limsup_{t\to d^-} \| u(t) \|_Y = +\infty$.

3. CONTINUOUS DEPENDENCE

We consider a sequence of integrodifferential equations

$$u'(t) + A_n(t)u(t) = \int_0^t g_n(t,s,u(s))ds + f_n(t) \tag{3.1}$$

$$u(0) = u_{0,n} \tag{3.2}$$

We assume that (A1)-(A5) hold and let $u(t)$ denote the solution of (1.1)-(1.2) on $[0,T_0]$. We seek conditions on (3.1)-(3.2) which guarantee that its solutions will converge to $u(t)$ as $n \to \infty$. In order to simplify the presentation, we discuss the special case where $A_n(t) \equiv A(t)$ does not depend on n. We assume that

(A7) f_n belongs to $C(0,\infty;X) \cap L^1_{loc}(0,\infty;Y)$ for each $n \geq 1$.

(A8) There is a nonempty open subset W of Y such that for each $n \geq 1$ the functions $g_n : \Delta \times W \to Y$ are continuous. Furthermore, for each $y \in W$ there is a constant $r > 0$ and a positive continuous function $b(t,s)$ defined on Δ such that if $B_Y(y;r) = \{z \in Y : \| z - y \|_Y \leq r\}$ then $B_Y(y;r) \subset W$ and

$$\| g_n(t,s,y_1) - g_n(t,s,y_2) \|_Y \leq b(t,s) \| y_1 - y_2 \|_Y$$

for all $y_1, y_2 \in B_Y(y;r)$, $(t,s) \in \Delta$ and all $n = 1, 2, 3, \ldots$.

THEOREM 3. In addition to the above assumptions, suppose that

(i) $\lim_{n\to\infty} \| g_n(t,s,y) - g(t,s,y) \|_Y = 0$ uniformly on compact subsets of $\Delta \times W$.

(ii) For each $T > 0$, $\lim_{n\to\infty} \int_0^T \| f_n(s) - f(s) \|_Y \, ds = 0$.

(iii) $\lim_{n\to\infty} \| u_{0,n} - u_0 \|_Y = 0$

Then there is a $T > 0$ ($T \leq T_0$) and an integer $N \geq 1$ such that for all $n \geq N$, equations (3.1)-(3.2) have a unique solution $u_n(t)$ on $[0,T]$ which satisfies

$$\lim_{n\to\infty} \max_{0\leq t\leq T} \| u_n(t) - u(t) \|_Y = 0$$

$$\lim_{n\to\infty} \int_0^T \| u_n'(t) - u'(t) \|_X \, dt = 0$$

4. AN APPLICATION

To illustrate the results of the previous section, we consider the Cauchy problem for the equation

$$\frac{\partial u}{\partial t} + a_1(x,t)\frac{\partial u}{\partial x} + a_0(x,t)u = \int_0^t g(t,s,u(x,s))ds + f(x,t) \qquad (4.1)$$

$$u(x,0) = u_0(x) \qquad (4.2)$$

where $x \in R$ and $t \in (0,\infty)$. We assume that $u = (u_1,\ldots,u_N)$, $g = (g_1,\ldots,g_N)$, $f = (f_1,\ldots,f_N)$ are N-vector functions and $a_1(x,t)$, $a_0(x,t)$ are $N \times N$ matrix functions. We will denote by D the differential operator $\partial/\partial x$.

Let $L^2 = L^2(R)$ be the Hilbert space of all square integrable complex-valued N-vector functions on R with inner product

$$(u,v) = \sum_{j=1}^{N} \int_{-\infty}^{+\infty} u_j(x)\,\overline{v_j(x)}\,dx$$

and corresponding norm $\| u \|_0 = \sqrt{(u,u)}$. Let $H^1 = H^1(R)$ denote the Sobolev space of N-vector functions u in L^2 having weak derivatives Du belonging to L^2. The norm on H^1 is given by $\| u \|_1 = (\| u \|_0^2 + \| Du \|_0^2)^{1/2}$.

For a given set E, we let $C_b(E)$ denote the Banach space of all $N \times N$ matrix-valued functions that are continuous and bounded on E. We let $C_b^1(R)$ denote the set of all $N \times N$ matrix-valued functions a such that a and Da belong to $C_b(R)$. These spaces are considered to be equipped with their corresponding supremum norms.

We make the following assumptions on equation (4.1).

(i) $a_1^*(x,t) = a_1(x,t)$, i.e., $a_1(x,t)$ is a Hermitian matrix.

(ii) $a_1 \in C(0,\infty;C_b^1(R))$

(iii) $a_0 \in C(0,\infty;C_b(R))$ and $Da_0 \in C_b(R \times [0,T])$ for each $T > 0$.

(iv) $f \in C(0,\infty;L^2) \cap L^1_{loc}(0,\infty;H^1)$

(v) $g = g(t,s,\eta) : \Delta \times R^N \to R^N$ is continuous with continuous first and second partial derivatives with respect to $\eta \in R^N$.

(vi) There are continuous functions $\nu : \Delta \to R^+$ and $\rho : R^+ \to R^+$, with ρ non-decreasing, such that $|g(t,s,\eta)| \le \nu(t,s)|\eta|\rho(|\eta|)$ for all (t,s,η) in $\Delta \times R^N$.

LEMMA 1. For each $u \in H^1$ let $g(t,s,u)(x) = g(t,s,u(x))$. Then $g : \Delta \times H^1 \to H^1$ is continuous and for each $(t,s) \in \Delta$ the mapping $u \to g(t,s,u)$ is locally Lipschitz continuous in u.

To define the family of negative generators associated with equation (4.1), we first let $A(t)$ denote the formal differential operator

$$A(t)u(x) = a_1(x,t)Du(x) + a_0(x,t)u(x)$$

Then the formal adjoint $A^*(t)$ is given by

$$A^*(t)v(x) = -D[a_1(x,t)v(x)] + a_0^*(x,t)v(x)$$

If u and ϕ are C^1 functions on R, with supp ϕ compact, it follows that

$$(A^*(t)\phi, u) = (\phi, A(t)u)$$

For each fixed $t \geq 0$ we define an operator A(t) in L^2 as follows. We say that $u \in D(A(t))$ if $u \in L^2$ and if there is a function $v \in L^2$ such that

$$(A^*(t)\phi, u) = (\phi, v)$$

for all C^1 functions ϕ with compact support in R. Put $A(t)u = v$. By a result of Friedrichs [2], A(t) is a closed, linear, densely defined operator in L^2 whose domain D(A(t)) contains the space H^1 and can be characterized as follows: $u \in D(A(t))$ if and only if $u \in L^2$ and there is a function $v \in L^2$ and a sequence $\{u_n\}$ in $C^\infty \cap L^2$ such that $A(t)u_n$ belongs to L^2 and $\| u - u_n \|_0 \to 0$, $\| v - A(t)u_n \|_0 \to 0$ as $n \to \infty$ (This is Theorem 1 of [2] plus a partition of unity argument). It is shown in [3] that A(t) is *quasi-accretive* in L^2, i.e.,

$$\| (A(t) + I)^{-n} \|_0 \leq (\lambda - \beta(t))^{-n} \qquad \text{for all } \lambda > \beta(t), \qquad n \geq 1$$

where

$$\beta(t) = \sup_{x \in R} |a_0(x,t) - \tfrac{1}{2} Da_1(x,t)|$$

This is equivalent to the fact that A(t) satisfies the energy inequality. It follows that A(t) is quasi-stable on each bounded interval [0,T] with stability index $\{1,\beta(t)\}$.

Assumptions (A2) and (A3), with $X = L^2$ and $Y = H^1$, are verified in [8]. Indeed, the operator S = S(t) can be chosen independent of t:

$$S = (1 - D^2)^{\frac{1}{2}}$$

and the operator B(t) is given by

$$B(t) = [S, M_1(t)]DS^{-1} + [S, M_0(t)]S^{-1}$$

where [S,T] = ST - TS denotes the commutator and $M_j(t)$ denotes multiplication by $a_j(x,t)$.

We conclude by Theorem 1 that the Cauchy problem (4.1)-(4.2) has a unique solution for each $u_0 \in H^1$.

REFERENCES

1. Dafermos, C. M., An Abstract Volterra Equation with Applications to Linear Viscoelasticity, *J. Diff. Eqs.* 7, 554-569 (1970).

2. Friedrichs, K. O., The Identity of Weak and Strong Extensions of Differential Operators, *Trans. Amer. Math. Soc.* 55 (1944), 132-151.

3. Friedrichs, K. O., Symmetric Positive Linear Differential Equations, *Comm. Pure Appl. Math.* 11 (1958), 333-418.

4. Goldstein, J. A., Semigroups and Second-Order Differential Equations, *J. Functional Analysis* 4, 50-70 (1969).

5. Goldstein, J. A., Time Dependent Hyperbolic Equations, *J. Functional Analysis* 4, 31-49 (1969).

6. Kato, T., Integration of the Equation of Evolution in a Banach Space, *J. Math. Soc. Japan*, 5, No. 2, July, 1953, 208-234.

7. Kato, T., On Linear Differential Equations in Banach Spaces, *Comm. Pure Appl. Math.* 9 (1956), 479-486.

8. Kato, T., Linear Evolution Equations of "Hyperbolic" Type, *J. Fac. Sci. Univ. Tokyo*, Sec. I. Vol 17 (1970), 241-258.

9. Kato, T., Linear Evolution Equations of "Hyperbolic" Type, II, *J. Math. Soc. Japan*, 25 (1973), 648-666.

10. Webb, G. F., An Abstract Semilinear Volterra Integrodifferential Equation, *Proc. Amer. Math. Soc.* 69, No. 2, May, 1978, 225-260.

11. Webb, G. F., Abstract Volterra Integrodifferential Equations and a Class of Reaction-Diffusion Equations, *Volterra Equations*, Lecture Notes in Mathematics, (1979) Vol. 737, pp. 295-303, Springer-Verlag, New York.

INTEGRABILITY OF RESOLVENTS OF VOLTERRA INTEGRODIFFERENTIAL EQUATIONS

G. Samuel Jordan

Department of Mathematics
University of Tennessee
Knoxville, Tennessee

1. INTRODUCTION

The linear Volterra integrodifferential system

$$x'(t) = Ax(t) + \int_0^t B(t - s)x(s)ds + f(t), \qquad 0 \leq t < \infty \tag{1.1}$$

$$x(0) = x_0$$

and perturbations of it have been studied extensively; see, for example, [2-6] and their references. Here, x(t) and f(t) are vector functions with n components and

$$\left\{\begin{array}{l} a = [a_{ij}] \text{ is a real constant } n \times n \text{ matrix} \\ B(t) = [b_{ij}(t)] \text{ is a real } n \times n \text{ matrix}, \qquad B(t) \in L^1(0,\infty) \end{array}\right. \tag{H_1}$$

It is well known that under mild assumptions the solution x(t) may be written as

$$x(t) = R(t)x_0 + \int_0^t R(t - s)f(s)ds, \qquad 0 \leq t < \infty \tag{1.2}$$

where the resolvent R(t) is the n × n matrix determined by

$$R'(t) = AR(t) + \int_0^t B(t - s)R(s)ds, \qquad 0 \le t < \infty$$
$$R(0) = I$$

(I is the identity matrix). The resolvent R(t) also occurs in "variation of constants" formulae (see, for example, [3]) which solve certain nonlinear perturbed forms of (1.1).

It is clear from (1.2) that it is useful to find conditions on B(t) which imply that R(t) is integrable on $(0,\infty)$. For then properties of f(t) such as boundedness, convergence at $+\infty$, and integrability on $(0,\infty)$ are propagated to the convolution term $\int_0^t R(t - s)f(s)ds$. Grossman and Miller [4] have proved that if $B(t) \in L^1(0,\infty)$, then $R(t) \in L^1(0,\infty)$ if and only if

$$\det[zI - A - B^*(z)] \neq 0 \qquad (\text{Re } z \ge 0) \tag{1.3}$$

where

$$B^*(z) = \int_0^\infty e^{-zt}B(t)dt$$

is the Laplace transform of B(t).

Although condition (1.3) is very useful, it is sometimes difficult to verify since the function on the left-hand side of (1.3) may be transcendental. For a graphical approach to determining whether (1.3) is satisfied, see [7]. Another approach is due to F. Brauer [2] who used condition (1.3) to establish the following result for the scalar case of (1.1):

THEOREM A. Let $n = 1$ and assume (H_1) holds. In addition, assume B(t) is continuous and of one sign on $[0,\infty)$ and that

$$T = \int_0^\infty t|B(t)|dt < \infty \tag{1.4}$$

If $A + \int_0^\infty B(t)dt \ge 0$, then R(t) is not integrable. If $\int_0^\infty B(t)dt > 0$ and $A + \int_0^\infty B(t)dt < 0$, then R(t) is integrable. If $\int_0^\infty B(t)dt < 0$ and $A + \int_0^\infty B(t)dt < 0$, then R(t) is integrable provided the moment T is sufficiently small.

Our primary purpose here is to present some sufficient conditions for determining whether R(t) is integrable when $n > 1$. The results are in the same spirit as those of Theorem A; moreover, when specialized to the case $n = 1$, they yield a strengthened version of Theorem A. Detailed proofs and further discussion may be found in [5]. (The results in [2] and [5] are stated in terms of the uniform asymptotic stability of the zero solution of (1.1) with $f(t) \equiv 0$; however, the uniform asymptotic stability of that solution is equivalent to the integrability of R(t) [6].)

2. RESULTS AND DISCUSSION

Let

$$\alpha_{ij} = \int_0^\infty b_{ij}(t)dt \qquad (i,\ j = 1,\ldots,n)$$

$$\beta_{ij} = \int_0^\infty |b_{ij}(t)|dt \qquad (i,\ j = 1,\ldots,n)$$

$$R_i = \sum_{j=1}^{n} (|a_{ij}| + \beta_{ij}) \qquad (i = 1,\ldots,n) \tag{2.1}$$

The first result gives sufficient conditions for integrability of R(t) in terms of these quantities and the elements of the matrix A. We shall also need the inequalities

$$\left\{\begin{array}{l} |a_{ii} + \alpha_{ii}|\ |a_{kk} + \alpha_{kk}| > \sum_{j=1}^{n}{}' |a_{ij} + \alpha_{ij}| \sum_{j=1}^{n}{}' |a_{kj} + \alpha_{kj}| \\ (i \neq k;\ i,\ k = 1,\ldots,n) \end{array}\right. \tag{H_2}$$

where a prime on a summation symbol indicates that the term $j = i$ is omitted. When $n = 1$, sums with primes have value zero and (H_2) is interpreted as $|a_{11} + \alpha_{11}| > 0$.

THEOREM 1. Let (H_1) and (H_2) hold. The resolvent R(t) is integrable if for each $i = 1,\ldots,n$ either of the following conditions holds:

$$a_{ii} < 0, \qquad |a_{ii}| \geq \sum_{j=1}^{n}{}' |a_{ij}| + \sum_{j=1}^{n} \beta_{ij} \tag{H_3}$$

$$\begin{cases} a_{ii} + \alpha_{ii} < 0, \quad T_i \equiv \int_0^\infty t|b_{ii}(t)|dt < \infty \\ T_i \leq (|a_{ii} + \alpha_{ii}| - \sum_{j=1}^{n}{}' (|a_{ij}| + \beta_{ij}))/R_i \end{cases} \tag{H_4}$$

The conclusion of Theorem 1 may or may not be true if equality is permitted in (H_2); see [5] for relevant examples.

By (H_2) $a_{ii} + \alpha_{ii} \neq 0$ for $i = 1,\dots,n$. It is assumed explicitly in (H_4) that $a_{ii} + \alpha_{ii} < 0$. This inequality also follows from (H_2), (H_3), and the obvious inequality $|\alpha_{ii}| \leq \beta_{ii}$. Thus, in Theorem 1 $a_{ii} + \alpha_{ii} < 0$ for each i. In the next theorem we consider the case $a_{ii} + \alpha_{ii} > 0$ for each i.

THEOREM 2. Let (H_1) and (H_2) hold and assume $a_{ii} + \alpha_{ii} > 0$ $(i = 1,\dots,n)$. If n is odd, then the resolvent R(t) is not integrable. If n is even, then R(t) may or may not be integrable.

Theorems 1 and 2 are proved by showing that condition (1.3) does or does not hold; a result of A. Brauer [1] on the location of eigenvalues of matrices plays a key role in the arguments. Both theorems remain true if the order of all subscripts is reversed in (2.1), (H_2), (H_3), and (H_4).

In the important scalar case of (1.1) the limiting case of (H_2), $a_{11} + \alpha_{11} = A + \int_0^\infty B(t)dt = 0$, implies that R(t) is not integrable (cf. Theorem A). This fact follows from the failure of (1.3) at $z = 0$: $(z - A - B^*(z))|_{z=0} = -a_{11} - \alpha_{11} = 0$. Combining Theorems 1 and 2 with this observation yields

THEOREM 3. Let $n = 1$ and assume (H_1) holds. If $A + \int_0^\infty B(t)dt \geq 0$, then R(t) is not integrable. If $A \leq \int_0^\infty B(t)dt < -A$, then R(t) is integrable. If the moment T defined in (1.4) is sufficiently small and either $\int_0^\infty B(t)dt < A < 0$ or $0 \leq A < -\int_0^\infty B(t)dt$, then R(t) is integrable.

This result obviously differs from Theorem A in that B(t) is assumed to be continuous and of one sign on $[0,\infty)$ in Theorem A. These assumptions are natural in the context of population models which motivate the results in [2]. More importantly, Theorem 3 imposes no moment condition when $A \leq \int_0^\infty B(t)dt < -A$; whereas, Theorem A requires a moment condition whenever both $A + \int_0^\infty B(t)dt$ and $\int_0^\infty B(t)dt$ are negative. An example given in [2] shows, however, that R(t) may be integrable when both these quantities are negative even if the moment T is large. One simply takes A = -1, $B(t) = (-1/a)\exp(-t/a)$ where a > 0; clearly, $\int_0^\infty B(t)dt = -1$ and T = a. Since $A = \int_0^\infty B(t)dt$ here, Theorem 3 applies for any moment T = a > 0.

We conclude with an example which indicates the usefulness of our results. Let A be a 2 × 2 matrix with $a_{11} = a_{22} = -3$, $a_{12} = a_{21} = 1/2$, and let B(t) be a 2 × 2 matrix with $b_{11}(t) = e^{-t}$, $b_{12}(t) = \int_t^\infty (e^{-v}/v)dv$, $b_{21}(t) = 1/2$ $(0 \leq t < 1)$, $b_{21}(t) = 0$ $(1 \leq t < \infty)$, and $b_{22}(t) = -e^{-t}/2$. Clearly, $B(t) \in L^1(0,\infty)$ and $\alpha_{11} = \beta_{11} = \alpha_{12} = \beta_{12} = 1$, $\alpha_{21} = \beta_{21} = -\alpha_{22} = \beta_{22} = 1/2$. Thus, (H_2) holds and (H_3) is satisfied for i = 1, 2 (for i = 2 (H_4) also holds). By Theorem 1 the resolvent R(t) associated with A and B(t) is integrable. The relative ease with which Theorem 1 may be applied in this example contrasts sharply with the difficulty of establishing directly the necessary and sufficient condition (1.3), which in this case becomes

$$[z + 3 - 1/(z + 1)][z + 3 - 1/(2z + 2)] + [((\log(z + 1))/z) + 1/2][((e^{-z} - 1)/z) - 1/2] \neq 0 \quad (\mathrm{Re}\ z \geq 0)$$

ACKNOWLEDGMENT

Partially supported by National Science Foundation Grant MCS-7903358.

REFERENCES

1. A. Brauer, Limits for the characteristic roots of a matrix, II, *Duke Math. J.* 14 (1947), 21-26.

2. F. Brauer, Asymptotic stability of a class of integro-differential equations, *J. Differential Equations* 28 (1978), 180-188.

3. S. I. Grossman and R. K. Miller, Perturbation theory for Volterra integrodifferential systems, *J. Differential Equations* 8 (1970), 457-474.

4. S. I. Grossman and R. K. Miller, Nonlinear Volterra integrodifferential systems with L^1-kernels, *J. Differential Equations* 13 (1973), 551-556.

5. G. S. Jordan, Asymptotic stability of a class of integrodifferential systems, *J. Differential Equations* 31 (1979), 359-365.

6. R. K. Miller, Asymptotic stability properties of linear Volterra integrodifferential equations, *J. Differential Equations* 10 (1971), 485-506.

7. R. K. Miller, Asymptotic stability and perturbations for linear Volterra integrodifferential systems, *Delay and Functional Differential Equations and their Applications*, pp. 257-268, Academic Press, New York, 1972.

NONLINEAR RETARDED PERTURBATION OF A LINEAR EVOLUTION SYSTEM

James H. Lightbourne, III*

Mathematics Department
Pan American University
Edinburgh, Texas

1. INTRODUCTION

Let X be a Banach space with norm $|\cdot|$; let $r > 0$ and $C = C([-r,0]; X)$ be the Banach space of continuous functions $\phi : [-r,0] \to X$ with norm $|\phi| = \max \{|\phi(\theta)| : \theta \in [-r,0]\}$. Let $W = \{W(t,s) : 0 \leq s \leq t \leq c\}$ be a continuous linear evolution system on X (i.e., for $0 \leq s \leq t \leq c$, $W(t,s)$ is a linear operator on X, $W(t,t) = I$ the identity on X, $W(t,s)\, W(s,u) = W(t,u)$ for $0 \leq u \leq s \leq t \leq c$, and for each $x \in X$ the map $(t,s) \to W(t,s)x$ is continuous), $\Omega \subset C$, and $F : [0,\infty) \times \Omega \to X$ be continuous. For $\sigma \geq 0$, we say that

$$
\begin{aligned}
x(t) &= W(t,\sigma)\phi(0) + \int_\sigma^t W(t,s)F(s,x_s)ds, \qquad t \geq \sigma \\
x_\sigma &= \phi \in \Omega
\end{aligned}
\tag{*}
$$

has a (local) solution at (σ,ϕ) provided there exist $b > 0$ and a continuous function $x : [-r + s, s + b] \to X$ which satisfies (*) with $x_t \in \Omega$ for $t \in [\sigma, \sigma + b]$. We define $x_t \in C$ by $x_t(\theta) = x(t + \theta)$ for $-r \leq \theta \leq 0$.

**Current affiliation*: Department of Mathematics, West Virginia University, Morgantown, West Virginia

As is essentially noted in Hale [3], $x : [-r + \sigma, \sigma + b] \to X$ is a solution to (*) if and only if for $t \in [\sigma, \sigma + b]$

$$x_t(\theta) = \begin{cases} \phi(t - \sigma + \theta) & -r \leq \theta \leq \sigma - t \\ W(t + \theta, \sigma)\phi(0) + \int_\sigma^{t+\theta} W(t + \theta, s)F(s,x_s)ds & \sigma - t \leq \theta \leq 0 \end{cases}$$

In this paper we exploit this observation and concentrate on the variation of parameters equation for (*):

$$[u(t)]\theta = \begin{cases} \phi(t + \theta - \sigma) & -r \leq \theta \leq \sigma - t \\ W(t + \theta, \sigma)\phi(0) + \int_\sigma^{t+\theta} W(t + \theta, s)F(s,u(s))ds & \\ \qquad \sigma - t \leq \theta \leq 0 \end{cases} \tag{**}$$

where $(t,\theta) \in [\sigma,b] \times [-r,0]$. Note that if $x : [-r + \sigma, \sigma + b] \to X$ is defined by

$$x(t) = \begin{cases} u(\sigma)t & \sigma - r \leq t \leq \sigma \\ u(t)\sigma & \sigma \leq t \end{cases}$$

then $x_t = u(t)$ for $t \in [\sigma, \sigma + b]$ and x satisfies (*) with $x_\sigma = \phi$. In fact, the primary objective of this paper is to give approximate solutions to (**) and to indicate a situation in which it is advantageous to consider (**) rather than (*). More precisely, we will establish existence of (local) solutions to (*) when $\Omega \subset X$ is closed. The criteria we obtain is the retarded differential equations analog of the criteria for ordinary differential equations used by several authors (see e.g., Martin [6] and [7] and references cited therein).

It should be emphasized that the "Nagumo" criteria we obtain is from considering (**) rather than (*). Although many other questions can be treated using (**), we shall limit our discussion. The reader is referred to Fitzgibbon [2], Travis and Webb [9], and Webb [10] for further discussions of (*) and examples applying the abstract results to partial functional differential equations. For related papers on existence on closed sets for retarded differential equations, the reader is referred to Leela and Moauro [4], Seifert [8], and Lightbourne [5].

In Section 2 we establish existence of approximate solutions to (**) and in Section 3 give a compactness condition on the composition of T and F which ensures existence of solutions to (*).

2. APPROXIMATE SOLUTIONS

The construction of the approximate solutions given in this section is motivated by the similar construction found in Lightbourne [5], which is influenced by the ideas found in Martin [7] and Webb [11]. See also Amann [1].

Let $\phi \in \Omega$ and for simplicity we will suppose $\sigma = 0$. Since F is continuous and W is a continuous linear evolution on X, there exist positive numbers $M \geq 1$, b, and R such that if $(t,\Psi) \in [0,b] \times \Omega$ with $|\Psi - \phi| \leq R$ then $|F(t,\Psi)| \leq M$, $|W(t,s)| \leq M$ for $0 \leq s \leq t \leq b$, and if $|\Psi| \leq bM(M + 1)$ then $|S(t)\phi - \phi + \Psi| \leq R$ for all $t \in [0,b]$. For $h \geq 0$, $S(h) : \mathcal{C} \to \mathcal{C}$ is defined by

$$[S(h)]\theta = \begin{cases} \phi(h + \theta) & -r \leq \theta \leq -h \\ W(h + \theta, 0)\phi(0) & -h \leq \theta \leq 0 \end{cases}$$

DEFINITION. The pair $\{v,(t_i)_0^N\}$ is said to be an ε-approximate solution to (**) on $[0,b]$ provided $v : [0,b] \to \mathcal{C}$ and $(t_i)_0^N$ is a partition of $[0,b]$ with $t_{i+1} - t_i \leq \varepsilon$ satisfying

(ε1) $v(0) = \phi$, $v(t_i) \in \Omega$ and $|v(t_i) - \phi| \leq R$ for all $1 \leq i \leq N$, and if $t \in [t_i, t_{i+1})$ then $v(t)$ is defined by

$$v(t)\theta = \begin{cases} [v(t_i)](t - t_i + \theta) & -r \leq \theta \leq t_i - t \\ W(t + \theta, t_i)v(t_i)0 + (t + \theta - t_i)W(t + \theta, t_i)F(t_i, v(t_i)) & \\ \qquad t_i - t \leq \theta \leq 0 \end{cases}$$

(ε2) $|v(t_{i+1}) - v(t_{i+1}^-)| \leq \varepsilon\,(t_{i+1} - t_i)$ where $v(t_{i+1}^-) = \lim_{t \to t_{i+1}^-} v(t)$; and

(ε3) if $|\theta - \lambda| \leq t_{i+1} - t_i$ and $0 \leq h \leq t_{i+1} - t_i$ then $|v(t_i)\theta - v(t_i)\lambda| \leq \varepsilon$ and

$$|W(t_i + h, t_i)v(t_i)\theta - v(t_i)\lambda| \leq \varepsilon$$

LEMMA. Let $\{v, (t_i)_0^N\}$ be an ε-approximate solution for (**) on [0,b] and for $0 \le i \le N$ let $F_i = F(t_i, v(t_i))$. For $0 \le k \le N$ we associate with $\{v, (t_i)_0^N\}$ the function $w^k : [t_k, b] \to C$ defined as follows. Let $t \in [t_k, b]$ and $k \le i \le N - 1$ such that $t \in [t_i, t_{i+1}]$. Then for $-r \le \theta \le t_k - t$ define

$$w^k(t)\theta = v(t_k)(t + \theta - t_k)$$

for $k + 1 \le l \le i$ and $t_{l-1} - t \le \theta \le t_l - t$ define

$$w^k(t)\theta = W(t - \theta, t_k)v(t_k)0 + \sum_{j=k}^{l-2} (t_{j+1} - t_j)W(t + \theta, t_j)F_j + (t - t_{l-1} + \theta)W(t + \theta, t_{l-1})F_{l-1}$$

and for $t_i - t \le \theta \le 0$ define

$$w^k(t)\theta = W(t + \theta, t_k)v(t_k)0 + \sum_{j=k}^{i-1} (t_{j+1} - t_j)W(t + \theta, t_j)F_j + (t + \theta - t_i)W(t + \theta, t_i)F_i$$

Then $w^k : [t_k, b] \to C$ is continuous and for $t \in [t_i, t_{i+1})$,

$$|w^k(t) - v(t)| \le \varepsilon M(t_i - t_k)$$

(Note: We interpret $\sum_{j=p}^{q} (\ ) = 0$ when $p > q$).

Proof. That $w^k : [t_k, b] \to C$ and is continuous is easily seen; that

$$|w^k(t) - v(t)| \le \varepsilon M(t_i - t_k)$$

is also easily seen once we observe that $v(t)$ can be written in the following form. For each $0 \le i \le N$ define $H_i \in C$ by $H_i(\theta) = v(t_i)\theta - v(t_i -)\theta$.

Then if $t \in [t_i, t_{i+1})$ and $0 \le k \le i$, we have for $-r \le \theta \le t_k - t$

$$v(t)\theta = v(t_k)(t + \theta - t_k) + \sum_{j=k+1}^{i} H_j(t - t_j + \theta)$$

for $k + 1 \le l \le i$ and $t_{l-1} - t \le \theta \le t_l - t$

$$v(t)\theta = W(t + \theta, t_k)v(t_k)0 + \sum_{j=k}^{l-2} (t_{j+1} - t_j)W(t + \theta, t_j)F_j$$
$$+ (t - t_{l-1} + \theta)W(t + \theta, t_{l-1})F_{l-1} + \sum_{j=l}^{i} H_j(t - t_j + \theta)$$
$$+ \sum_{j=k+1}^{l-1} W(t + \theta, t_j)H_j(0)$$

and for $t_i - t \le \theta \le 0$

$$v(t)\theta = W(t + \theta, t_k)v(t_k)0 + \sum_{j=k}^{i-1} (t_{j+1} - t_j)W(t + \theta, t_j)F_j$$
$$+ (t + \theta - t_i)W(t + \theta, t_i)F_i + \sum_{j=k+1}^{i} W(t + \theta, t_j)H_j(0)$$

(Here again we interpret $\sum_{j=p}^{q} (\ ) = 0$ when $p > q$.)

This expression for $v(t)$ can be established by an inductive argument which we will merely indicate by verifying the case $k = i - 1$. For $t \in [t_i, t_{i+1})$

$$v(t)\theta = \begin{cases} v(t_i)(t - t_i + \theta) & -r \le \theta \le t_i - t \\ W(t + \theta, t_i)v(t_i)0 + (t + \theta - t_i)W(t + \theta, t_i)F_i & t_i - t \le \theta \le 0 \end{cases}$$

$$= \begin{cases} v(t_i -)(t - t_i + \theta) + H_i(t - t_i + \theta) & -r \le \theta \le t_i - t \\ W(t + \theta, t_i)v(t_i -)0 + (t + \theta - t_i)W(t + \theta, t_i)F_i \\ \quad + W(t + \theta, t_i)H_i(0) & t_i - t \le \theta \le 0 \end{cases}$$

$$= \begin{cases} v(t_{i-1})(t - t_{i-1} + \theta) + H_i(t - t_i + \theta) & -r \le \theta \le t_{i-1} - t \\ W(t + \theta, t_{i-1})v(t_{i-1})0 + (t + \theta - t_i)W(t + \theta, t_{i-1})F_{i-1} \\ \quad + H_i(t - t_i + \theta) & t_{i-1} - t \le \theta \le t_i - t \\ W(t + \theta, t_{i-1})v(t_{i-1})0 + (t_i - t_{i-1})W(t + \theta, t_{i-1})F_{i-1} \\ \quad + (t + \theta - t_i)W(t + \theta, t_i)F_i + W(t + \theta, t_i)H_i(0) \\ \quad t_i - t \le \theta \le 0 \end{cases}$$

This completes our indication of proof.

We will frequently refer to the function $G_h : [0,\infty) \times C \to C$ defined for $h \ge 0$ by

$$[G_h(t,\phi)]\theta = \begin{cases} \phi(h + \theta) & -r \le \theta \le -h \\ W(t + h + \theta, t)\phi(0) + (h + \theta)W(t + h + \theta, t)F(t,\phi) \\ \quad -h \le \theta \le 0 \end{cases}$$

Also, for $\overline{\Psi} \in C$ define $|\overline{\Psi}; \Omega| = \inf \{|\overline{\Psi} - \Psi| : \Psi \in \Omega\}$.

PROPOSITION. Let $\Omega \subset C$ be closed, $F : [0,\infty) \times \Omega \to X$ be continuous, and $\lim_{h \to 0^+} \inf |G_h(t,\Psi); \Omega|/h = 0$ for each $(t,\Psi) \in [0,\infty) \times \Omega$. Suppose $\phi \in \Omega$ and b, R, and M are as above. Let $\{\varepsilon_n\}_1^\infty \subset (0,1)$ with $\lim \varepsilon_n = 0$. Then for each n there exists $N = N(n)$ and an ε_n-approximate solution $\{u_n, (t_i^n)_{i=0}^N\}$ to (**) on $[0,b]$.

REMARK. The critical conditions for existence on closed sets is the "tangential" assumption:

$$\lim_{h \to 0^+} \inf |G_h(t,\Psi); \Omega|/h = 0 \text{ for all } (t,\Psi) \in [0,\infty) \times \Omega \qquad (T_1)$$

As with an ordinary differential equation this condition is always satisfied when Ω is open; actually, our results are valid under the weaker condition that Ω be locally closed which includes Ω open. Also, if $\Lambda \subset X$ is closed and convex, $\Omega = \{\phi \in C : \phi(\theta) \in \Lambda \text{ for all } -r \leq \theta \leq 0\}$, $W(t,s) : \Lambda \to \Lambda$, and

$$\lim_{h\to 0^+} \inf |\Psi(0) + hF(t,\Psi); \Omega|/h \text{ for all } (t,\Psi) \in [0,\infty) \times \Omega \qquad (T_2)$$

then (T_1) is satisfied. The reader is referred to Martin [7] for examples in which $W(t,s) : \Lambda \to \Lambda$ and although ordinary differential equations are treated, based on the examples in [7] one can construct examples in which (T_2) is satisfied.

Proof of Proposition. We give an inductive construction of $\{u_n, (t_i^n)_{i=0}^N\}$. Suppose u_n is defined on $[0, t_i^n]$ and $t_i^n < b$; let δ_i^n be the maximum number satisfying

(i) $\delta_i^n \in [0, \varepsilon_n]$ with $t_i^n + \delta_i^n \leq b$;

(ii) if $|\theta - \lambda| \leq \delta_i^n$ and $h \in [0, \delta_i^n]$ then $|u_n(t_i^n)\theta - u_n(t_i^n)\lambda| \leq \varepsilon_n$ and $|W(t_i^n + h, t_i^n)u_n(t_i^n)\theta - u_n(t_i^n)\lambda| \leq \varepsilon_n$;

(iii) $|G_{\delta_i^n}(t_i^n, u_n(t_i^n)); \Omega| \leq \delta_i^n \varepsilon_n/2$.

The suppositions on F and W ensure that $\delta_i^n > 0$. Define $t_{i+1}^n = t_i^n + \delta_i^n$ and for $t \in [t_i^n, t_{i+1}^n)$ define $u_n(t) \in C$ by

$$[u_n(t)\theta] = \begin{cases} [u_n(t_i^n)](t - t_i^n + \theta) & -r \leq \theta \leq t_i^n - t \\ W(t + \theta, t_i^n)u_n(t_i^n)0 + (t + \theta - t_i^n)W(t + \theta, t_i^n)F(t_i^n, u_n(t_i^n)) & t_i^n - t \leq \theta \leq 0 \end{cases}$$

Note that $u_n(t_{i+1}^{n-}) = G_{\delta_i^n}(t_i^n, u_n(t_i^n))$ and thus by (iii) we have $|u_n(t_{i+1}^{n-}); \Omega| \leq \delta_i^n \varepsilon_n/2$. We choose $u_n(t_{i+1}^{n-}) \in \Omega$ such that $|u_n(t_{i+1}^{n-}) - u_n(t_{i+1}^n)| \leq (t_{i+1}^n - t_i^n)\, \varepsilon_n$. That $\{u_n, (t_j^n)_{j=0}^{i+1}\}$ is an ε_n-approximate solution to (**) on $[0, t_{i+1}^n]$ follows once we establish that $|u_n(t_{i+1}^n) - \phi| \leq R$. To see that this is the case, let $w_n = w_n^0$ be associated with the ε_n-approximate solution u_n for $k = 0$ as in the Lemma and define $\Psi \in C$ by

$$\Psi(\theta) = \begin{cases} u_n(t_{i+1}^n)\theta - w_n(t_{i+1}^n)\theta & -r \le \theta \le -t_{i+1}^n \\ u_n(t_{i+1}^n)\theta - W(t_{i+1}^n + \theta, 0)\phi(0) & -t_{i+1}^n \le \theta \le 0 \end{cases}$$

We have

$$\begin{aligned} |u_n(t_{i+1}^n) - \phi| &= |u_n(t_{i+1}^n) - w_n(t_{i+1}^n) + w_n(t_{i+1}^n) - \phi| \\ &= |\Psi + S(t_{i+1}^n)\phi - \phi| \end{aligned}$$

and by the definition of w_n and the Lemma

$$\begin{aligned} |\Psi| &\le |u_n(t_{i+1}^n) - w_n(t_{i+1}^n)| + M^2 t_{i+1}^n \\ &\le |u_n(t_{i+1}^n) - u_n(t_{i+1}^{n-})| + |u_n(t_{i+1}^{n-}) - w_n(t_{i+1}^n)| + M^2 t_{i+1}^n \\ &\le \varepsilon_n(t_{i+1}^n - t_i^n) + \varepsilon_n M t_i^n + M^2 t_{i+1}^n \\ &\le \varepsilon_n M(t_{i+1}^n - t_i^n) + \varepsilon_n M t_i^n + M^2 t_{i+1}^n \\ &= (\varepsilon_n M + M^2) t_{i+1}^n \le bM(M + 1) \end{aligned}$$

By the selection of b and M we have $|u_n(t_{i+1}^n) - \phi| = |\Psi + S(t_{i+1}^n)\phi - \phi| \le R$.

The proof of the Proposition will be complete once we establish that there exists a positive integer $N = N(n)$ such that $t_N^n = b$. Suppose for contradiction that $t_i^n < b$ for all i and $\lim_{i\to\infty} t_i^n = b^* < b$. Let w_n^k be associated with u_n as in the Lemma and recall by the Lemma $|w_n^k(t_i^n) - u_n(t_i^n)| \le \varepsilon_n(t_i^n - t_k^n) \le \varepsilon_n(b^* - t_k^n)$. One can show $\{w_n^k(t_i^n)\}_{i=k}^\infty$ is a Cauchy sequence; consequently $\lim_{i\to\infty} u_n(t_i^n)$ exists and since Ω is closed we have $\lim_{i\to\infty} u_n(t_i^n) = u_\infty \in \Omega$. The compactness of $\{u_n(t_i^n)\}_{i=1}^\infty \cup \{u_\infty\}$, the continuity of F, and the properties of W yield that there exist $\delta > 0$ such that if $h \in [0,\delta]$ then h may replace δ_i^n in (i). Since the family $\{u_n(t_i^n)\} \cup \{u_\infty\}$ is compact, the family is equicontinuous and therefore we may assume if $h \in [0,\delta]$ then h may replace δ_i^n in (ii). Also, there exists $h_0 \in [0,\delta]$ such that $|G_{h_0}(b^*,u_\infty); \Omega| \leq h_0\varepsilon_n/3$. Since for sufficiently large i we have $\delta_i^n < h_0$ and δ_i^n is the largest number such that (i)-(iii) is satisfied, it must be that $|G_{h_0}(t_i^n, u_n(t_i^n)); \Omega| \geq h_0\varepsilon_n/2$ for sufficiently large i. But this is impossible since

$$h_0\varepsilon_n/3 \geq |G_{h_0}(b^*,u_\infty); \Omega| = \lim_{i\to\infty} |G_{h_0}(t_i^n, u_n(t_i^n)); \Omega| \geq h_0\varepsilon_n/2$$

We conclude that there exists $N = N(n)$ such that $t_N^n = b$, which completes the proof of the Proposition.

3. EXISTENCE

Although one may obtain existence results under other assumptions on W and F, we shall only give a criteria involving compactness on the composition of the evolution system W and F. In particular, suppose

for any positive numbers $s_1 < s_2$, b, R there exists a compact set $K = K(s_1,s_2,b,R) \subset X$ such that $W(s_2,s_1)F(t,x) \in K$ if $t \in [0,b]$ and $|x| \leq R$. (E)

THEOREM. Suppose $\Omega \subset C$, $F : [0,\infty) \times \Omega \to X$ is continuous, W is a linear evolution system on X and (E) holds. Then for each $(\sigma,\phi) \in [0,\infty) \times \Omega$ there exists a (local) solution x(t) to (*) with $x_t \in \Omega$ for $t \geq \sigma$ if and only if $\lim_{h\to 0^+} \inf |G_h(t,\Psi); \Omega|/h = 0$ for all $(t,\Psi) \in [0,\infty) \times \Omega$.

Proof. We first show necessity. If $(\sigma,\phi) \in [0,\infty) \times \Omega$ then $x : [-r + \sigma, \sigma + t] \to X$ is a solution to (*) if and only if x_t is a solution of (**). Consequently,

$$\begin{aligned}\lim_{h\to 0^+} \inf |G_h(\sigma,\phi); \Omega|/h &\leq \lim_{h\to 0^+} |G_h(\sigma,\phi) - x_{\sigma+h}|/h \\ &= \lim_{h\to 0^+} \max_{-h\leq\theta\leq 0} |(h + \theta)W(\sigma + h + \theta, \sigma)F(\sigma,\phi) \\ &\qquad - \int_\sigma^{\sigma+h+\theta} W(\sigma + h + \theta, s)F(s,x_s)ds|/h\end{aligned}$$

$$= \lim_{h \to 0^+} \max_{-h \le \theta \le 0} \int_\sigma^{\sigma+h+\theta} |W(\sigma + h + \theta, \sigma)F(\sigma,\phi) - W(\sigma + h + \theta, s)F(s,x_s)|\,ds/h$$

$$= 0$$

To establish sufficiency, we suppose for convenience that $\sigma = 0$ and $\phi \in \Omega$ and show that there exist $b > 0$ and a continuous function $u : [0,b] \to \Omega$ such that $\lim_{n\to\infty} u_n = u$ uniformly on $[0,b]$, where $\{u_n, (t_i^n)\}$ are ε_n-approximate solutions to (**). It readily follows that u satisfies (**) and as noted in the introduction, the theorem follows.

For each ε_n-approximate solution $\{u_n, (t_i^n)\}$ let w_n be the associated function for $k = 0$ as defined in the Lemma. Using condition (E) and the continuity of the map $(t,s) \to W(t,s)x$, one can show that $\{w_n\}_{n=1}^\infty$ is an equicontinuous family. To apply the Ascoli-Arzela theorem to the family $\{w_n\}$ we must establish that for each $t \in [0,b]$ we have $\{w_n(t)\}_{n=1}^\infty \subset C$ is precompact; i.e., for each $t \in [0,b]$ the family $\{w_n(t)\}_{n=1}^\infty \subset C$ is equicontinuous and for each $\theta \in [-r,0]$ the set $\{w_n(t)\theta\}_{n=1}^\infty \subset X$ is precompact. Again, one can show that this is the case using condition (E) and the continuity of the map $(t,s) \to W(t,s)x$. Thus by the Ascoli-Arzela theorem $\{w_n\}_{n=1}^\infty$ has a subsequence uniformly convergent on $[0,b]$. Relabelling if necessary, there exists a continuous function $u : [0,b] \to C$ such that $\lim_{n\to\infty} w_n = u$ uniformly on $[0,b]$ and from the Lemma we have that $\lim_{n\to\infty} u_n = u$ uniformly on $[0,b]$. For $t \in [t_i^n, t_{i+1}^n)$,

$$u_n(t)\theta - u_n(t_i^n)\theta = \begin{cases} [u_n(t_i^n)](t - t_i^n + \theta) - [u_n(t_i^n)]\theta & -r \le \theta \le t_i^n - t \\ W(t + \theta, t_i^n)u_n(t_i^n)0 \\ \quad + (t + \theta - t_i^n)W(t + \theta, t_i^n)F(t_i^n, u_n(t_i^n)) \\ \quad - u_n(t_i^n)\theta & t_i^n - t \le \theta \le 0 \end{cases}$$

By (ε3) in the definition of ε-approximate solution, we obtain

$$|u_n(t)\theta - u_n(t_i^n)\theta| \le \varepsilon_n \text{ for } -r \le \theta \le t_i^n - t \text{ and for } t_i^n - t \le \theta \le 0$$

$$|W(t + \theta, t_i^n)u_n(t_i^n)0 + (t + \theta - t_i^n)W(t + \theta, t_i^n)F(t_i^n,u_n(t_i^n)) - u_n(t_i^n)\theta|$$

$$\leq \varepsilon_n M^2 + |W(t + \theta,\ t_i^n)u_n(t_i^n)0 - u_n(t_i^n)\theta|$$

$$\leq \varepsilon_n M^2 + \varepsilon_n$$

Thus,

$$u(t) = \lim_{n\to\infty} u_n(t) = \lim_{n\to\infty} u_n(\gamma_n(t)) \in \Omega$$

uniformly on $[0,b]$, where if $t \in [t_i^n, t_{i+1}^n)$ then $\gamma_n(t) = t_i^n$ and $\gamma_n(b) = b$. By the continuity of F and the compactness of $\{(t, u(t)) : 0 \leq t \leq b\}$ we have $\lim_{n\to\infty} F(\gamma_n(t), u_n(\gamma_n(t))) = F(t,u(t))$ uniformly on $[0,b]$, from which it follows that for $t \in [0,b]$ and $\theta \in [-r,0]$

$$\begin{aligned} u(t)\theta &= \lim_{n\to\infty} w_n(t)\theta \\ &= \begin{cases} \phi(t + \theta) & -r \leq \theta \leq -t \\ W(t + \theta,\ 0)\phi(0) + \int_0^{t+\theta} \lim_{n\to\infty} W(t + \theta,\ \gamma_n(s))F(\gamma_n(s), \\ \qquad u_n(\gamma_n(s))ds & -t \leq \theta \leq 0 \end{cases} \\ &= \begin{cases} \phi(t + \theta) & -r \leq \theta \leq -t \\ W(t + \theta,\ 0)\phi(0) + \int_0^{t+\theta} W(t + \theta,\ s)F(s,\ u(s))ds & -t \leq \theta \leq 0 \end{cases} \end{aligned}$$

As indicated above, the assertions of the Theorem follow.

REFERENCES

1. H. Amann, Invariant sets and existence theorems for semilinear parabolic and elliptic systems, *J. Math. Anal. Appl.* 65 (1978), 432-467.

2. W. E. Fitzgibbon, Semilinear functional differential equations in Banach space, *J. Differential Equations* 29 (1978), 1-14.

3. J. Hale, *Functional Differential Equations*, Springer-Verlag, New York, 1971.

4. S. Leela and V. Moauro, Existence of solutions in a closed set for delay differential equations in Banach space, *J. Nonlinear Analysis* 2 (1978), 47-58.

5. J. H. Lightbourne, III, Function space flow invariance for functional differential equations of retarded type, *Proc. Amer. Math. Soc.*, 77 (1979), 91-98.

6. R. H. Martin, Jr., Differential equations on closed subsets of a Banach space, *Trans. Amer. Math. Soc.* 179 (1973), 399-414.

7. R. H. Martin, Jr., Invariant sets for perturbed semigroups of linear operators, *Ann. Mat. Pura. Appl.* 105 (1975), 221-239.

8. G. Seifert, Positively invariant closed sets for systems of delay differential equations, *J. Differential Equations* 22 (1976), 292-304.

9. C. C. Travis and G. F. Webb, Existence and stability for partial functional differential equations, *Trans. Amer. Math. Soc.* 200 (1974), 395-418.

10. G. F. Webb, Asymptotic stability for abstract nonlinear functional differential equations, *Proc. Amer. Math. Soc.* 54 (1976), 225-230.

11. G. F. Webb, Continuous nonlinear perturbations of linear accretive operators in Banach space, *J. Functional Analysis* 10 (1972), 191-203.

A SOLVABILITY CONDITION FOR SYSTEMS OF FUNCTIONAL DIFFERENTIAL EQUATIONS

John W. Neuberger

Department of Mathematics
North Texas State University
Denton, Texas

1. INTRODUCTION

This note contains a summary of recent results concerning square integrable solutions to linear inhomogeneous systems. The general setting not only includes a variety of systems of ordinary and partial differential equations but also includes many ordinary functional differential equations (OFDE) with both advanced and retarded arguments as well as corresponding partial functional differential equations (PFDE).

A sample OFDE is the following: Given r, s, g, continuous real valued functions with r, s bounded, g square integrable, and $\alpha, \beta \in R$, find y such that y, y' are square integrable on R and

$$y''(t) + r(t)y(t + \alpha) + s(t)y(t + \beta) = g(t), \quad t \in R \tag{1}$$

For $\alpha\beta < 0$ this is an OFDE with both advanced and retarded arguments. Necessary and sufficient conditions on g are sought for solvability.

This note is based mainly upon [5,6]. In [6] the added problem of boundary conditions is considered. See [2,3] for general background on OFDE.

2. ABSTRACT SETTING

Denote by each of H and K a Hilbert space and by H' a closed subspace of H. Denote by P the orthogonal projection of H onto H'. Denote by B a member of L(H,K) with the special property that BB* = I, the identity on K. Given $g \in K$, we seek $x \in H$ such that

$$BPx = g \tag{2}$$

It is of course possible that $R(BP) \neq K$ and hence there is for some $g \in K$ no solution x to the above problem. It is certainly possible that $R(BP) \neq \overline{R(BP)}$, i.e., that R(BP) is not closed. Finally $\overline{R(BP)}$ may well not be equal to K. The following theorems shed some light on such things. Denote BPB* by M. From [5] there are the following two theorems:

THEOREM 1. If $g \in K$, then $g \in \overline{R(BP)}$, if and only if $\lim_{n\to\infty} (I - M)^n g = 0$.

THEOREM 2. If $g \in \overline{R(BP)}$, then $g \in R(BP)$ if and only if

$$\lim_{n\to\infty} PB^* [I + (I - M) + \ldots + (I-M)^{n-1}]g$$

exists.

Denote by Φ a normalized spectral family for M (cf [7], p. 272). From [6] there is:

THEOREM 3. If $g \in \overline{R(BP)}$, then $g \in R(BP)$, if

$$\lim_{\delta\to 0+} \int_\delta^1 (1/j)d\langle\Phi g,g\rangle$$

exists.

Here j denotes the identity function on R and $\langle\Phi g,g\rangle$ denotes the function f on [0,1] so that $f(t) = \langle\Phi(t)g,g\rangle$, $0 \leq t \leq 1$.

A direct application of Theorem 3 would seem to require a construction of the spectral family Φ. M is commonly not compact and it seems difficult to construct Φ even though in applications M is often explicitly given. A purpose of the following theorem is to take a step in the direction of obtaining more easily verifiable conditions than the one of Theorem 3.

Denote by $h : [0,1] \to R$ so that $h(0) = 0$, $h(t) = 1/t$, $0 < t \leq 1$. Denote by $\{B_n\}_{n=i}^{\infty}$ the sequence of Bernstein polynomials for h (cf [4]):

$$B_n(t) = \sum_{i=0}^{n} \binom{n}{i} t^i (1 - t)^{n-i} h(i/n), \qquad 0 \leq t \leq 1$$

THEOREM 4. If $g \in \overline{R(BP)}$, then $g \in R(BP)$ provided that

$$\{\langle B_n(M)g,g\rangle\}_{n=1}^{\infty}$$

has a bounded infinite subsequence.

3. EXAMPLES

Following are examples of problems that can be expressed in the form (2). The first example is (1), the second a system of PFDE of rather general type.

EXAMPLE 1. Denote $L_2(R)^4$ by H, $L_2(R)^2$ by K. Denote by A the element of L(H,K) so that if

$$x = \begin{pmatrix} p \\ q \\ u \\ v \end{pmatrix} \in H, \qquad \text{then } Ax = \begin{pmatrix} q - u \\ \\ v + rp_\alpha + sp_\beta \end{pmatrix}$$

Where, if $f \in L_2(R)$ and $\gamma \in R$ then $f_\gamma(t) = f(t + \gamma)$ for almost all t.

Denote by P the orthogonal projection on H so that if

$$x = \begin{pmatrix} p \\ q \\ u \\ v \end{pmatrix} \in H$$

then

$$Px = \begin{pmatrix} y \\ y' \\ w \\ w' \end{pmatrix} \quad \text{so that} \quad \left\| \begin{pmatrix} p \\ q \\ u \\ v \end{pmatrix} - \begin{pmatrix} y \\ y' \\ w \\ w' \end{pmatrix} \right\|^2$$

is minimum where y', w' denote L_2 generalized derivatives (cf [1]).

Define also $G_0 = \binom{0}{g} \in K$. Note that AA^* is symmetric and that $(AA^*)^{-1}$ exists and belongs to $L(K,K)$ since $AA^* \geq I$. Define $B = (AA^*)^{-\frac{1}{2}}A$, $G = (AA^*)^{-\frac{1}{2}}G_0$. If

$$BPx = G$$

then for some $x \in H$, it follows that

$$Px = \begin{pmatrix} y \\ y' \\ w \\ w' \end{pmatrix}$$

for some $y, w \in L_2(R)$ and so

$$B\begin{pmatrix} y \\ y' \\ w \\ w' \end{pmatrix} = G$$

Hence

$$A\begin{pmatrix} y \\ y' \\ w \\ w' \end{pmatrix} = G_0$$

i.e.,

$$\begin{cases} y' - w = 0 \\ w' + ry_\alpha + sy_\beta = g \end{cases}$$

i.e.,

$$y'' + ry_\alpha + sy_\beta = g$$

i.e., (1) is satisfied.

Denote by Q the orthogonal projection on $L_2(R)^2$ so that if $\binom{f}{h} \in L_2(R)^2$, then $Q\binom{f}{g} = \binom{y}{y'}$ so that $\| \binom{f}{g} - \binom{y}{y'} \|^2$ is minimized. From [5] one has that the element y of $L_2(R)$ which accomplishes this is given by

$$y(t) = (e^t/2) \int_t^\infty e^{-j}(f - g) + (e^{-t}/2) \int_\infty^t e^j(f + g), \qquad t \in R$$

Note that if $\begin{pmatrix} p \\ q \\ r \\ s \end{pmatrix} \in H$, then $P \begin{pmatrix} p \\ q \\ r \\ s \end{pmatrix}$ is given by

$$\begin{pmatrix} Q\binom{p}{q} \\ Q\binom{r}{s} \end{pmatrix}$$

EXAMPLE 2. For a FPDE, denote by Ω an open subset of R^2 and define $H = L_2(\Omega)^6$, $K = L_2(\Omega)^2$. Denote by Q^{ij}, $i, j = 1, 2$, members of $L(L_2(\Omega), L_2(\Omega))$ and denote by

$$E = \begin{pmatrix} E_{11} & E_{12} & E_{13} & E_{14} \\ E_{21} & E_{22} & E_{23} & E_{24} \end{pmatrix}$$

a member of $L(R^4, R^2)$ so that $(EE^*)^{-1}$ exists. Define $A : H \to K$ so that if

$$x = \begin{pmatrix} a \\ b \\ c \\ d \\ e \\ f \end{pmatrix} \in H \text{ then}$$

$$(Ax)(p) = E\begin{pmatrix} b(p) \\ c(p) \\ e(p) \\ f(p) \end{pmatrix} + \begin{pmatrix} (Q^{11}a)(p) + (Q^{12}d)(p) \\ (Q^{21}a)(p) + (Q^{22}d)(p) \end{pmatrix}$$

for almost all $p \in \Omega$.

Note that $(AA^*)^{-1}$ exists. Define P to be the orthogonal projection on H so that if

$$x = \begin{pmatrix} a \\ b \\ c \\ d \\ e \\ f \end{pmatrix} \in H$$

then Px is the nearest member of H to x of the form

$$\begin{pmatrix} u \\ u_1 \\ u_2 \\ v \\ v_1 \\ v_2 \end{pmatrix}$$

where, for example, u_1 denotes the L_2 generalized partial derivative of u in the first argument of u. One may compute u by solving the Euler equation (with appropriate boundary conditions) for the minimization problem for u:

$$\left\| \begin{pmatrix} u \\ u_1 \\ u_2 \end{pmatrix} - \begin{pmatrix} a \\ b \\ c \end{pmatrix} \right\|^2$$

A similar comment holds for v.

Now suppose that $G \equiv \binom{g}{h} \in K$. It follows that the problem of finding $x \in H$ so that

$$APx = G \tag{3}$$

is equivalent to finding $u, v \in L_2(\Omega)$ so that

(i) the L_2 generalized derivatives u_1, u_2, v_1, v_2 exist

$$\text{(ii)} \quad \begin{cases} E_{11}u_1 + E_{12}u_2 + E_{13}v_1 + E_{14}v_2 + Q^{11}u + Q^{12}v = g \\ E_{21}u + E_{22}u_2 + E_{23}v_1 + E_{24}v_2 + Q^{21}u + Q^{22}v = h \end{cases} \tag{4}$$

Defining $B = (AA^*)^{-\frac{1}{2}}A$, $G = (AA^*)^{-\frac{1}{2}}\binom{g}{h}$, one has that solving each of both (3) and (4) is equivalent to finding $x \in H$ such that

$$BPx = G$$

Note in passing that if the action of each of Q^{ij}, i, j = 1, 2, is multiplication by a bounded continuous real-valued function on Ω, then (4) is a conventional system of two partial differential equations. In its full generality, (4) seems to have been studied very little. Perhaps the theorems in this note provide a starting place.

4. PROOF OF THEOREM 4

Suppose that $\{\langle B_n(M)g,g\rangle\}_{n=1}^{\infty}$ has a bounded subsequence but that

$$\lim_{\delta\to 0+} \int_\delta^1 (1/j)d\langle\Phi g,g\rangle$$

does not exist. Denote by $\{n_i\}_{i=1}^{\infty}$ an increasing sequence of positive integers and by c a positive number so that $\langle B_{n_i}(M)g,g\rangle \le c$, $i = 1, 2, \ldots$. Denote by δ a positive number so that

$$\int_\delta^1 (1/j)d\langle\Phi g,g\rangle > c$$

Now $\{B_{n_i}\}_{i=1}^{\infty}$ converges uniformly to $1/j$ on $[\delta,1]$. Hence

$$\begin{aligned} c < \int_\delta^1 (1/j)d\langle\Phi g,g\rangle &= \lim_{i\to\infty} \int_\delta^1 B_{n_i}\; d\langle\Phi g,g\rangle \\ &\le \liminf_{i\to\infty} \int_0^1 B_{n_i}\; d\langle\Phi g,g\rangle = \liminf_{i\to\infty} \langle\int_0^1 B_{n_i}\; d\Phi g,g\rangle \\ &= \liminf_{i\to\infty} \langle B_{n_i}(M)g,g\rangle \le c \end{aligned}$$

a contradiction. The above calculation uses the fact that for each positive integer n, $\langle M^n g,g\rangle = \langle\int_0^1 j^n d\Phi g,g\rangle = \int_0^1 j^n d\langle\Phi g,g\rangle$ and so $\int_0^1 B_{n_i}\; d\langle\Phi g,g\rangle = \langle\int_0^1 B_{n_i}\; d\Phi g,g\rangle$, $i = 1, 2, \ldots$.

Hence the conclusion follows using Theorem 3.

This work is part of a project on iterative methods for both linear and nonlinear systems of differential equations. Parts of this development have led to computer methods concerning boundary value problems for systems of nonlinear partial differential equations. There is good reason to feel that the theory of the present note may be implemented to solve functional differential equations numerically.

REFERENCES

1. S. Agmon, *Lectures on Elliptic Boundary Value Problems*, D. van Nostrand, 1965.

2. Bellman and Cooke, *Differential Difference Equations*, Academic Press, New York, 1963.

3. J. K. Hale, *Functional Differential Equations*, Springer-Verlag, New York, 1971.

4. G. G. Lorentz, *Bernstein Polynomials*, Univ. of Toronto, 1953.

5. J. W. Neuberger, Square Integrable Solutions to Linear Inhomogeneous Systems, *J. Diff. Equations*, 27 (1978), 144-152.

6. J. W. Neuberger, Boundary Value Problems for Linear Systems, *Proc. Royal Society Edinburgh*, V. 83A (1979), 297-302.

7. J. W. Neuberger, A Type-independent Interative Method for Systems of Nonlinear Partial Differential Equations: Application to the Problem of Transonic Flow, *Computers and Mathematics with Applications*, 6 (1980), 67-78.

8. F. Reisz and B. Sz-Nagy, *Functional Analysis*, Ungar, New York, 1955.

CONVERGENCE OF SOLUTIONS OF DIFFERENTIAL EQUATIONS WITH INFINITE DELAY

Mary E. Parrott*

Department of Mathematics
Memphis State University
Memphis, Tennessee

1. INTRODUCTION

We consider the limiting behavior as $t \to \infty$ of solutions of the functional differential equation (FDE) with infinite (unbounded) delay

$$x'(t) = f(t,x_t) \tag{1.1}$$

where $f : [0,\infty) \times B \to R^n$ is continuous, B is a space of functions mapping $(-\infty,0]$ into R^n, and $x_t \in B$ is defined by $x_t(s) = x(t + s)$, $-\infty < s \leq 0$. We say that a solution $x(\cdot)$ of (1.1) converges if $\lim_{t\to\infty} x(t)$ exists. The question of convergence of solutions of (1.1) is related to the questions of stability and asymptotic stability of solutions, but is appropriate at times when asymptotic stability considerations are not. This is the situation, for example, when any constant function is a solution of (1.1).

Both the form of f in (1.1) and the choice of the underlying space B are seen to influence the methods of studying the problem of convergence of solutions, as well as asymptotic stability, of infinite delay equations. The purpose of this paper is to use Liapunov-Razumikhin techniques to give

*Current affiliation: Department of Mathematics, University of South Florida, Tampa, Florida

sufficient conditions for the convergence of solutions of (1.1) when B = BC, the bounded continuous functions mapping $(-\infty,0]$ into R^n, and then indicate that by modifying the techniques we obtain a convergence result when $B = C_\gamma$, the space of continuous functions, ϕ, mapping $(-\infty,0]$ into R^n, such that the limit, $\lim_{s\to-\infty} e^{\gamma s}\phi(s)$ $(\gamma > 0)$, exists.

2. CONVERGENCE OF SOLUTIONS WITH UNDERLYING SPACE BC (I)

We first consider the FDE with infinite delay

$$x'(t) = f(t,x_t) \tag{2.1}$$

where $f : [0,\infty) \times BC \to R^n$ is continuous. BC is a Banach space with norm, $\|\phi\| = \sup_{s\le 0}|\phi(s)|$, for $\phi \in BC$. Driver [2] laid the foundation for a qualitative theory of infinite delay equations with underlying space BC by giving conditions which insure the existence, uniqueness, continuous dependence of solutions on initial data, and continuability of solutions of the system (2.1). We will assume henceforth that there exists a solution to (2.1) and that f satisfies the necessary conditions for bounded solutions of (2.1) to be continuable.

DEFINITION. A Liapunov function V (on $R \times R^n$) is a continuous nonnegative scalar function mapping $R \times R^n$ to $[0,\infty)$.

Our first convergence result was motivated by Driver's asymptotic stability theorem using Liapunov functions [2, Theorem 7]. Another motivating factor came from Kato [5] who studied the convergence properties of Liapunov functions (along solutions to finite-delay equations) in order to obtain information about the convergence of solutions. Our methods in this section hinge on being able to find a Liapunov function $V = V(t,x)$ and (for each t) an interval $I_t = [\lambda(t),t]$ where

(i) $\lambda(t) \to \infty$ monotonically as $t \to \infty$;

(ii) $\lim_{t\to\infty} [\sup_{u \in [\lambda(t),t]} V(u,x(u))]$ exists; and

(iii) there is a sequence $\{t_n\} \to \infty$ as $n \to \infty$ such that $\lim_{n\to\infty} V(t_n,x(t_n)) = \lim_{t\to\infty} [\sup_{u \in [\lambda(t),t]} V(u,x(u))]$.

In order to implement (ii) above we give the following result. (For the proof of this result we refer the reader to [6, Lemmas 2.4, 2.5 and Theorem 2.4].)

DEFINITION. If $V : R \times R^n \to [0,\infty)$ is a Liapunov function and $x(\cdot;t,\phi)$ is a solution of (2.1) with $x_t = \phi \in BC$, define

$$V'_{(2.1)}(t,\phi) = \overline{\lim_{h\to 0^+}} \frac{1}{h} [V(t + h, x(t + h; t, \phi)) - V(t,\phi(0))]$$

THEOREM 2.1. Suppose there exist a Lipaunov function $V = V(t,x)$ on $R \times R^n$, a continuous function $\lambda : [0,\infty) \to R$ such that $\lambda(t) \leq t$ for all $t \geq 0$ and $\lambda(t) \to \infty$ monotonically as $t \to \infty$, and a measurable function $p : [0,\infty) \to R$ such that $p \in L^1[0,\infty)$. If

$$V'_{(2.1)}(t,\phi) \leq p(t)$$

whenever $t \geq 0$, $\phi \in BC$ and $\sup_{s\in J_t} V(t + s, \phi(s)) = V(t, \phi(0))$, where $J_t = [\lambda(t) - t,0]$, then for any solution $x(\cdot)$ of (2.1) that is defined in the future,

$$\lim_{t\to\infty} [\sup_{s\in J_t} V(t + s, x(t + s))]$$

exists.

REMARK 2.1. If $p(t) \equiv 0$ in Theorem 2.1 then it can be shown [6, Corollary 2.5.1] that $\sup_{s\in J_t} V(t + s, x(t + s))$ is a nonincreasing function of t. This conclusion also follows from a comparison result of Driver [2, Lemma 1].

REMARK 2.2. If all of the conditions of Theorem 2.1 are satisfied and in addition $V(t,x)$ is radially unbounded (that is, $V(t,x) \to \infty$ as $|x| \to \infty$

uniformly for $t \in [0,\infty)$) then all solutions of (2.1) are bounded in the future and the conclusion of Theorem 2.1 is: for any solution of (2.1), $\lim_{t\to\infty} [\sup_{s\in J_t} V(t + s, x(t + s))]$ exists [6, Corollary 2.4.1].

We now state and prove our main convergence result for BC.

THEOREM 2.2. Suppose there exists a Liapunov function $V = V(t,x)$ on $R \times R^n$ with the property that $\lim_{t\to\infty} x(t) = 0$ whenever $\lim_{t\to\infty} V(t,x(t)) = 0$ where $x(\cdot)$ is a solution of (2.1) defined in the future. Suppose also that there exist a continuous real-valued function h such that $h(u) > u$ for $u > 0$, measurable functions p_i, $i = 1, 2$, $p_i : [0,\infty) \to R$ such that $p_i \in L^1[0,\infty)$, and a continuous function $\lambda : [0,\infty) \to R$ such that $\lambda(t) \leq t$ and $\lambda(t) \to \infty$ monotonically as $t \to \infty$. If

$$V'_{(2.1)}(t,\phi) \leq p_1(t)$$

whenever $t \geq 0$, $\phi \in BC$ and $\sup_{s\in J_t} V(t + s, \phi(s)) = V(t,\phi(0))$, where $J_t = [\lambda(t) - t,0]$, and if for every $\varepsilon > 0$ there is a $\delta = \delta(\varepsilon) > 0$ and a $K = K(\varepsilon) > 0$ such that

$$V'_{(2.1)}(t,\phi) \leq -\delta|f(t,\phi)| + Kp_2(t)$$

whenever $t \geq 0$, $\phi \in BC$, $\sup_{s\in J_t} V(t + s, \phi(s)) \leq 2\varepsilon$, $V(t,\phi(0)) \geq \varepsilon$ and $V(t + s, \phi(s)) < h(V(t,\phi(0)))$ for $s \in J_t$, then for any solution $x(\cdot)$ of (2.1) that is defined in the future, $\lim_{t\to\infty} x(t)$ exists.

Proof. Let V, h, p_i, $i = 1, 2$, and λ be as described in the theorem. Let $x(\cdot)$ be a solution of (2.1) that is defined in the future. Since the conditions of Theorem 2.1 are satisfied, $\lim_{t\to\infty} [\sup_{s\in J_t} V(t + s, x(t + s))] = \alpha < \infty$. It can be shown [6, Lemma 2.3] that there is a sequence $\{t_n\} \to \infty$ as $n \to \infty$ such that $V(t_n,x(t_n)) \to \alpha$, and that the hypotheses of the theorem imply that $\lim_{t\to\infty} V(t,x(t)) = \alpha$ [6, Theorem 2.5]. If $\alpha = 0$ this implies $\lim_{t\to\infty} x(t) = 0$ by the hypothesis on V, so assume $\alpha > 0$. Choose ε such that $0 < \varepsilon < \alpha < 2\varepsilon$, and let $\eta(\varepsilon) = \min_{\varepsilon\leq u\leq 2\varepsilon} [h(u) - u]$. Choose T sufficiently large that $|\sup_{s\in J_t} V(t+s, x(t+s)) - \alpha| < \eta(\varepsilon)/2$, $\sup_{s\in J_t} V(t+s, x(t+s)) \le 2\varepsilon$, $V(t, x(t)) \ge \varepsilon$ and $|\alpha - V(t,x(t))| < \eta(\varepsilon)/2$ for $t \ge T$. Then, for $t \ge T$,

$$\begin{aligned}
&\sup_{s\in J_t} V(t+s, x(t+s)) - V(t, x(t)) \\
&\quad = |\sup_{s\in J_t} V(t+s, x(t+s)) - V(t, x(t))| \\
&\quad \le |\sup_{s\in J_t} V(t+s, x(t+s)) - \alpha| + |\alpha - V(t, x(t))| \\
&\quad \le \frac{\eta(\varepsilon)}{2} + \frac{\eta(\varepsilon)}{2} = \eta(\varepsilon) = \min_{\varepsilon\le u\le 2\varepsilon} [h(u) - u] \\
&\quad \le h(V(t, x(t))) - V(t, x(t))
\end{aligned}$$

(since for $t > T$, $\varepsilon \le V(t, x(t)) \le 2\varepsilon$). But this yields

$$\sup_{s\in J_t} V(t+s, x(t+s)) < h(V(t, x(t))) \quad \text{for} \quad t \ge T$$

which implies (by the hypotheses of the theorem)

$$V'_{(2.1)}(t,x_t) \le -\delta|f(t,x_t)| + Kp_2(t)$$

Hence, for any sequence $\{t_n\}$ such that $\{t_n\} \to \infty$ as $n \to \infty$,

$$\begin{aligned}
V(t_n, x(t_n)) - V(t_m, x(t_m)) &\le \int_{t_m}^{t_n} [-\delta|f(t, x_t)| + Kp_2(t)]dt \\
&\le -\delta\left|\int_{t_m}^{t_n} f(t, x_t)dt\right| + K\left|\int_{t_m}^{t_n} p_2(t)dt\right|
\end{aligned}$$

Thus,

$$V(t_n, x(t_n)) - V(t_m, x(t_m)) - K\int_{t_m}^{t_n} |p_2(t)|dt \le -\delta|x(t_n) - x(t_m)| \le 0$$

But, $\lim_{m,n\to\infty} (V(t_n, x(t_n)) - V(t_m, x(t_m))) = 0$ since $\lim_{t\to\infty} V(t,x(t))$ exists. Also, $\lim_{m,n\to\infty} \int_{t_m}^{t_n} |p_2(t)|dt = 0$ since $p_2 \in L^1[0,\infty)$. Hence, $|x(t_n) - x(t_m)| \to 0$

as $n,m \to \infty$ which implies $\{x(t_n)\}$ is a Cauchy sequence, which, in turn, implies $\lim_{t\to\infty} x(t)$ exists.

EXAMPLE 2.1. Consider the scalar equation

$$x'(t) = [-a(t)x(t) + b(t)x(t - r(t))]G(t,x_t) + p(t) \tag{2.2}$$

where $t \geq 0$, $a(t) \geq 0$, $|b(t)| \leq \theta a(t)$, $0 \leq \theta < 1$, and p is a continuous function such that $p \in L^1[0,\infty)$. Further, we assume $0 \leq r(t)$, $t - r(t) \to \infty$ as $t \to \infty$, $G(t,\phi) \geq 0$ for all $t \geq 0$ and $\phi \in BC$, and G satisfies a local Lipschitz condition in ϕ. We will show that if $x(\cdot)$ is any solution of (2.2), then $x(t) \to$ a constant as $t \to \infty$.

Let $V = x^2/2$ and $h(u) = q^2u$ where $q > 1$ is chosen so that $\theta q < 1$. If $t - r(t) \to \infty$ monotonically as $t \to \infty$ let $\lambda(t) = t - r(t)$; otherwise define $\lambda(t)$ by:

$$\lambda(t) = \inf \{u = s - r(s) : u \leq t \quad \text{for} \quad s \geq t\}$$

Then in either case $\lambda(t) \to \infty$ monotonically as $t \to \infty$. Let $\varepsilon > 0$ be given. For $t \geq 0$, $\phi \in BC$, it can readily be shown that

$$V'_{(2.2)}(t,\phi) \leq -a(t)[1 - \theta q]\sqrt{2\varepsilon}|\phi(0)|G(t,\phi) + p(t)$$

and

$$|f(t,\phi)| \leq a(t)[1 + \theta q]|\phi(0)|G(t,\phi) + p(t)$$

whenever $V(t, \phi(0)) \geq \varepsilon$ and $V(t + s, \phi(s)) < h(V(t, \phi(0)))$ for $s \in J_t = [\lambda(t) - t,0]$. Thus,

$$V'_{(2.2)}(t,\phi) \leq -\delta|f(t,\phi)| + Kp(t)$$

under the above conditions, where $\delta = \delta(\varepsilon) = ((1 - \theta q)/(1 + \theta q)) \sqrt{2\varepsilon}$ and $K = K(\varepsilon) = 1 + \delta$. Since $V = x^2/2$ is radially unbounded, all solutions of equation (2.2) are bounded in the future (cf. Remark 2.2), and hence all solutions of (2.2) are defined in the future. Therefore we can conclude from Theorem 2.2 that if $x(\cdot)$ is any solution of (2.2), $\lim_{t\to\infty} x(t)$ exists.

REMARK 2.3. Theorem 2.2 is not an asymptotic stability result. Zero is not in general a solution of equation (2.2); if, in equation (2.2), $p(t) \equiv 0$ and $G(t,\phi) = \int_{-t}^{0} [\phi(s) - \phi(0)]^2 ds$, then any constant function is a solution of (2.2) and asymptotic stability of the zero solution is impossible. For a discussion of the stability and asymptotic stability conclusions which can be drawn from Theorem 2.2, when appropriate, we refer the reader to [6, Chapter 3].

REMARK 2.4. If, in equation (2.2), $p(t) \equiv 0$ and $G(t,\phi) \equiv 1$, then the equation is a standard scalar example considered by Driver [2, Example 11] and others. However, in order to satisfy the usual requirement (in asymptotic stability theorems for FDEs) that $V'_{(2.1)}(t,\phi)$ be negative definite (for certain values of t and ϕ), one must ask that $a(t) \geq a > 0$. We can show asymptotic stability of the zero solutions of such equations with $a(t) \geq 0$ and $\int_0^\infty a(s)ds = \infty$ by applying Theorem 2.2, and then showing that all solutions converge to zero. (cf. [6, Proposition 3.1]).

It is often the case in applications that $\lambda(t)$ will be $t - r(t)$ where $r(t)$ is the "largest" delay appearing in the right-hand side of the equation. However, we can apply Theorem 2.2 to the equation

$$x'(t) = \frac{A}{t+1} x(t - \beta t) \tag{2.3}$$

where $t \geq 0$ and A is an $n \times n$ stable matrix, to show that all solutions of (2.3) tend to zero if $0 \leq \beta < 1 - \exp(-\lambda/(2\mu|B|\ |A|^2))$ where B is the positive definite matrix solution of $A^T B + BA = -I$ (the identity matrix) and λ^2 and μ^2 are the smallest and largest eigenvalues of B, respectively [6, Example 2.3]. In (2.3) $r(t)$ is chosen larger than βt.

3. CONVERGENCE OF SOLUTIONS WITH UNDERLYING SPACE BC (II)

It is natural to ask if Theorem 2.2 can be generalized so as to provide information about the convergence of solutions of an equation like

$$x'(t) = a(t)x(t) + \int_0^t b(t,s)x(s)ds \tag{3.1}$$

If, in Theorem 2.2, we do not require a $\lambda(t)$ which tends to ∞ as $t \to \infty$ to determine an interval over which the inequalities on $V'_{(2.1)}(t,\phi)$ are asked to hold, but rather ask that the inequalities hold for all $s \leq 0$, then the resulting statement is false. The scalar equation

$$x'(t) = -2x(t) + \sin(t)x(0) \tag{3.2}$$

with $V = |x|$, satisfies the modified statement of Theorem 2.2, but (3.2) can be solved explicitly to show that solutions do not tend to constants as $t \to \infty$. This example suggests that a different approach must be taken when considering infinite delay equations where the space of initial functions is BC, but the equations depend on the past over an interval whose left-hand endpoint is fixed. This is the situation in equation (3.1) as well as in more general Volterra integrodifferential systems. We now give a convergence result for BC (related to Theorem 2.2) which is applicable to certain of these Volterra equations. (For the proof we refer the reader to [6, Theorem 4.3].)

THEOREM 3.1. Suppose there exists a Liapunov function $V = V(t,x)$ on $R \times R^n$ with the property that $\lim_{t\to\infty} x(t) = 0$ whenever $\lim_{t\to\infty} V(t,x(t)) = 0$, where $x(\cdot)$ is a solution of (2.1) defined in the future. Suppose also that there exist measurable functions $p_i : [0,\infty) \to R$ such that $p_i \in L^1[0,\infty)$, $i = 1, 2$. Assume the following conditions are satisfied:

(i) There exists a $T_1 > 0$ such that

$$V'_{(2.1)}(t,\phi) \leq p_1(t)$$

whenever $\phi \in BC$, $\sup_{s\leq 0} V(t + s, \phi(s)) = V(t, \phi(0))$ and $t \geq T_1$, and

(ii) for every ε_1, ε_2, $0 < \varepsilon_1 < \varepsilon_2$, there is a $T = T(\varepsilon_1,\varepsilon_2)$, a $K = K(\varepsilon_1,\varepsilon_2)$ and a $\delta = \delta(\varepsilon_1,\varepsilon_2)$ such that

$$V'_{(2.1)}(t,\phi) \leq -\delta|f(t,\phi)| + Kp_2(t)$$

whenever $\phi \in BC$, $\sup_{s\leq 0} V(t + s, \phi(s)) \leq \varepsilon_2$, $V(t, \phi(0)) \geq \varepsilon_1$ and $t \geq T$.

If $x(\cdot)$ is a solution of (2.1) that is defined in the future, then $\lim_{t\to\infty} x(t)$ exists.

EXAMPLE 3.1. Consider the equation

$$x'(t) = a(t)Ax(t) + \int_0^t B(t,s)x(s)ds \tag{3.3}$$

where $t \geq 0$, A and B are $n \times n$ matrices, A is a stable matrix, B is locally integrable on $[0,\infty) \times [0,\infty)$, $a(t) \geq 0$, $\int_0^\infty a(s)ds = \infty$, and for every $\varepsilon > 0$ there is a $T > 0$ such that $\int_0^t |B(t,s)|ds \leq \varepsilon a(t)$ for $t \geq T$. In addition, assume that B satisfies the following conditions (in order for the right-hand side of (3.3) to be continuous):

$$\lim_{h\to 0} \int_0^t |B(t,s) - B(t + h, s)|ds = 0 \qquad \text{for each} \qquad t \geq 0 \tag{3.4}$$

and

$$\lim_{h\to 0} \int_t^{t+h} |B(t + h, s)|ds = 0 \qquad \text{for each} \qquad t \geq 0 \tag{3.5}$$

If

$$\lim_{t\to\infty} \int_0^t |B(t,s)|ds = 0 \tag{3.6}$$

then any solution of (3.3) tends to a constant as $t \to \infty$. In fact, all solutions of (3.3) tend to zero as $t \to \infty$.

Let $V(t,x) = x^TCx$, where C is the positive definite, symmetric matrix solution of the equation $A^TC + CA = -I$. Let λ^2 and μ^2 denote the smallest and largest eigenvalues of C, respectively. Then it can be shown (cf. [6, Example 4.2]) that for ε_1, ε_2 and θ given, $0 < \varepsilon_1 < \varepsilon_2$, $0 \leq \theta < 1$, and T chosen large enough that

$$\int_{-t}^0 |B(t, t + s)|ds \leq a(t) \left(\frac{\lambda}{\mu}\right)^2 \left(\frac{\varepsilon_1}{\varepsilon_2}\right) \left(\frac{\theta}{2}\right) \left(\frac{1}{|C|}\right)$$

$$\overset{\text{def}}{=} a(t)k(\varepsilon_1, \varepsilon_2) \qquad \text{for} \qquad t \geq T$$

$$V'_{(2.1)}(t,\phi) \leq -\delta|f(t,\phi)|$$

whenever $t \geq 0$, $\phi \in BC$, $\sup_{s \leq 0} V(t + s, \phi(s)) \leq \varepsilon_2$, $V(t, \phi(0)) \geq \varepsilon_1$ and $t \geq T$, where

$$\delta = \delta(\varepsilon_1, \varepsilon_2) = \frac{(1 - \theta)\varepsilon_1}{\frac{\sqrt{\varepsilon_2}\mu^2}{\lambda}[|A| + k(\varepsilon_1, \varepsilon_2)]}$$

Since V is radially unbounded and the conditions of Theorem 3.1 are satisfied we can conclude that if $x(\cdot)$ is any solution of (3.3) then $\lim_{t\to\infty} x(t)$ exists. The convergence of all solutions to zero follows from the boundedness of all solutions in the future and the conditions $\int_0^\infty a(s)ds = \infty$ and (3.6).

If, in equation (3.3), $a(t) \equiv 1$, then we have the type of Volterra integrodifferential equation considered, for example, by Grimmer and Seifert [3]. In [3, Theorem 5] Grimmer and Seifert show that all solutions of equation 3.3 (with $a(t) \equiv 1$) tend to zero as $t \to \infty$ under the given conditions (3.4) and (3.5), along with the conditions (replacing (3.6)):

There exists a constant $M > 0$ so that $\int_0^t |CB(t,s)|ds \leq M$ for all $t \geq 0$, and $2\mu M/\lambda < 1$; and (3.7)

$$\limsup_{T\to\infty} \{\int_0^{t-T} |B(t,s)|ds : t \geq T\} = 0 \tag{3.8}$$

4. CONVERGENCE OF SOLUTIONS WITH UNDERLYING SPACE C_γ

In some sense the choice of BC for the space of initial functions is undesirable as far as a qualitative theory of stability is concerned. For example, the zero solution of (2.1) can be asymptotically stable in R^n but not asymptotically stable in BC. Also, the set $\{x_t, t \geq 0\} \subset BC$ is not necessarily precompact, even if $x(\cdot)$ is a solution of (2.1) that is bounded in the future. In a recent paper Hale and Kato [4] have sought to remedy this situation by considering an axiomized space of initial functions, B. Prototype spaces satisfying the axioms in [4] are the so-called fading memory spaces considered, for example, by Coleman and Mizel in [1].

Another space which satisfies the axioms in [4] is the space C_γ, the space of continuous functions, ϕ, mapping $(-\infty,0]$ to R^n, such that the limit, $\lim_{s\to-\infty} e^{\gamma s}\phi(s)$ $(\gamma > 0)$, exists. (The space C_0 has some of the same undesirable features as BC.) C is a Banach space with norm, $\| \phi \| = \sup_{s\leq 0} e^{\gamma s}|\phi(s)|$, for $\phi \in C_\gamma$.

By modifying some of the techniques used in obtaining our convergence results for BC, Theorems 2.2 and 3.1, we are able to obtain a related convergence result for C_γ [6, Theorem 5.2]. A key idea (again, in the Razumikhin spirit) which is used in the development of this result is to find conditions which insure that the limit of an appropriate functional of V exists (see [6, Lemma 5.1, Corollary 5.1.1, and Lemma 5.3]). Statements and proofs of these results will follow in a subsequent paper.

REFERENCES

1. B. Coleman and V. Mizel, On the stability of solutions of functional-differential equations, *Arch. Rat. Mech. Anal.* 30 (1968), 173-196.

2. R. Driver, Existence and stability of solutions of a delay-differential system, *Arch. Rat. Mech. Anal.* 10 (1962), 401-426.

3. R. Grimmer and G. Seifert, Stability properties of Volterra integro-differential equations, *J. Differential Eqns.* 19 (1975), 142-166.

4. J. Hale and J. Kato, Phase space for retarded equations with infinite delay, *Funkcial Ekvac.* 21 (1978), 11-41.

5. J. Kato, On Liapunov-Razumikhin type theorems for functional differential equations, *Funkcial Ekvac.* 16 (1973), 225-239.

6. M. Parrott, Convergence of solutions of functional differential equations with infinite delay, Ph.D. thesis, Memphis State University, Memphis, TN.

PROPERTIES OF AN INTEGRAL EQUATION DESCRIBING ENZYME ACTIVITY

William Peterson

Department of Mathematics
West Virginia University
Morgantown, West Virginia

Frank Verhoff

Department of Chemical Engineering
West Virginia University
Morgantown, West Virginia

1. INTRODUCTION

Mathematical problems have their origin in many human activities. The integral equation to be discussed herein arose from the technology responsible for world-wide economic changes in the sweetener market. There are further strategic implications to the development of this technology. This is not to say that the solution of the integral equation greatly changed the course of this economic development, but it is interesting that such a technology as sugar could give rise to a variant of classical integral equations.

Sucrose, the dominant sweetener until four or five years ago, was obtained from two sources, sugar beets and sugar cane. Sugar from cane has always been inherently more economical to produce than that from beets. However, sugar cane can only be grown in the tropic regions of the world and in order for many countries to maintain an internal supply of sugar, the sugar beet industry was supported by several governments, e.g. United States, Australia, and the Soviet Union. This situation persisted until the middle of the 1970s when a third even more economical source of sweetener, corn, was technologically developed. With this development, there was an excess capacity for the production of sugar and the price of sugar plummented. Such sugar cane producing countries as Cuba, the Philippines, and the state of Hawaii have been economically depressed because of the decline in sugar prices. Further, the sugar beet industry in Ohio, Michigan, and elsewhere in the midwest of the U. S. is rapidly vanishing. The United States is abandoning its support of the sugar beet industry since it can be self-sufficient in sweetener by producing corn instead of beets. It will be several years until the consequences of this technology are fully effected.

For years a sugar, glucose, has been produced from corn and marketed, e.g., Karo syrup. However, this sugar is not very sweet although it is nutritionally as good as sucrose. Glucose was derived from corn by first wet milling the corn and then treating the corn starch in a large vessel with the enzyme amylose. The resulting solution of primarily glucose and some other oligosaccharides was then heated to destroy the amylose activity and the product was sold. It was also known that glucose could be converted by the enzyme glucose isomerase to the very sweet sugar, fructose, which is the principle component of honey. The chemistry of this conversion is shown below.

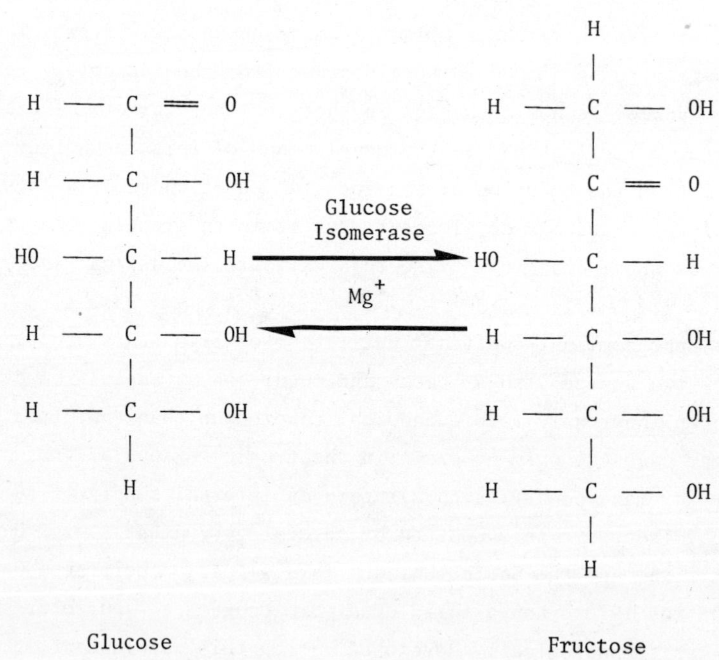

Although this is a relatively simple conversion, it was not economical because the glucose isomerase was too expensive to be put in a vessel with the glucose and destroyed after the conversion as was the case with amylose.

In the late 1960s and early 1970s efforts were made to immobilize the glucose isomerase on solid particles such that the glucose would be pumped past the particles and the conversion of glucose to fructose achieved. In this manner the glucose isomerase could be reused many times (more than 1000). Compared to the soluble form in a vessel, the process for the conversion of glucose to fructose then is economical. Although there is a much greater utilization of the glucose isomerase in the immobilized form, the immobilized enzyme still looses its activity over a period of time. Thus some compensation in the process must be made to keep the yield of fructose high. In some equipment the flow rate to the reactors is lowered as the enzyme activity decays; in others the temperature is raised to speed the reaction. This paper discusses a third alternative to maintaining the activity of the reactors by the continuous addition of fresh enzyme.

2. DEVELOPMENT OF THE EQUATION

The first assumption made in the analysis is that the decay rate of the enzyme is only a function of time and not a function of the pH, the glucose concentration, or any other factor. Enzyme decay rates are discussed by Laidler and Bunding [2]. Thus the activity of any mass of enzyme is only a function of the time it has been in the reactors; this is denoted a(t). In particular, an exponential decay and a linear decay will be considered.

We will first consider the case of two reactors in series. In a continuous operation of these two reactors, fresh enzyme is constantly added to the second reactor to compensate for the decay of enzyme in the two reactors. When the second reactor is full, the first reactor is taken off stream and a new column is added as the second reactor with the reactor just filled becoming the first reactor. When the new reactor is added to the system it may have some enzyme already in it if the reactor which was removed contained some enzyme activity. The goal of the analysis then is to find the enzyme addition rate.

After the start up period and the establishment of a cyclical steady state the enzyme addition rate function will be the same for each column. Additionally, the amount of enzyme initially in the reactor must be calculated. The total volume of the reactor is determined by the integral of the volumetric enzyme addition rate function f(t), over the time of the cycle plus the volume of enzyme initially in the reactor. If a(t) denotes the activity per unit volume for the enzyme as a function of time, then it is possible to compute the activity in each of the two columns at a given time. At time t, the activity in the column to which additions are being made will be given by the formula:

$$A_2(t) = \int_0^t a(t - s)f(s)ds + m_0a(t)$$

where

m_0 = volume of enzyme initially in the reactor,

f(t) = volumetric rate of enzyme addition,

a(t) = volumetric enzyme activity as a function of time in the reactor,

$A_2(t)$ = total enzyme activity in column 2 at time t.

Since the same enzyme addition rate was used to fill the first column, it is possible to calculate the enzyme activity in the first column at time t during the filling of the second column. Since the first column is full, we will integrate over the entire time of the column filling, i.e., ω. In addition, all the enzyme in the first column has been there for at least a time period t.

These concepts lead to the following equation for the activity in the first column.

$$A_1(t) = \int_0^\omega a(t + \omega - s)f(s)ds + m_0a(t + \omega)$$

where

$A_1(t)$ = total activity in the first column at time t during the filling of the second column,

ω = the time necessary to fill one of the columns.

For the reactors to maintain the necessary enzyme activity it is required that the sum of the activity in the two reactors be constant. So we have

$$T = A_1(t) + A_2(t)$$

$$= m_0\,[a(t) + a(t + \omega)] + \int_0^{\omega} a(t + \omega - s)f(s)ds$$

$$+ \int_0^{t} a(t - s)f(s)ds$$

where

T = total activity in both columns (could be a function of time).

This then is the type of integral equation to be investigated in this paper. It is interesting to note that this equation did not come from nor can it in general be reduced to a differential equation as have so many other applied integral equations. It should be noted that several other investigators have considered the enzyme decay from different points of view and have often used differential equations, e.g., Lee and Reilly [3].

3. PARTICULAR PROBLEMS

Similar to the two tank reactor, an (n + 1)-tank reactor can be described by

$$T(t) = m_0 \sum_{m=0}^{n} a(t + m\omega) + \sum_{m=1}^{n} \int_0^{\omega} a(t - s + m\omega)f(s)ds$$
$$+ \int_0^{t} a(t - s)f(s)ds \quad \text{for} \quad 0 \le t \le \omega \tag{1}$$

where

$$m_0 = [T(0) - \sum_{m=1}^{n} \int_0^{\omega} a(m\omega - s)f(s)ds] / \sum_{m=0}^{n} a(m\omega)$$

In the application of interest the activity function a(t) has zero order decay or first order decay.

For the case where $a(t) = Ee^{-kt}$, $E > 0$ and $k > 0$ equation (1) simplifies to

$$T(t) = T(0)e^{-kt} + \int_0^{t} Ee^{-k(t - s)}f(s)ds$$

an equivalent Volterra form with a unique solution on $L_2[0,\omega]$, assuming $T(t)$ in $L_2[0,\omega]$. See Tricomi [4] or Davis [1].

Assuming $T(t)$ is differentiable on $0 \leq t \leq \omega$ we have the solution of the form

$$f(t) = \frac{T'(t) + kT(t)}{E}$$

where

$$m_0 = [e^{k\omega n} - 1]/E(e^{k\omega(n+1)} - 1)$$

which will satisfy the physical constraints of the $(n + 1)$-tank problem provided that $T(t)$ has the property that $T'(t) + k\,T(t) \geq 0$ for $0 \leq t \leq \omega$ and $m_0 \geq 0$.

For the special case of constant output of the $(n + 1)$-tank problem (say $T(t) \equiv 1$) we have

$$f(t) = \frac{k}{E}$$

where

$$m_0 = \frac{1}{E(e^{k\omega} + 1)}$$

Note that from equation (1) that whenever

$$\frac{\sum_{m=0}^{n} a(t + m\omega)}{\sum_{m=0}^{n} a(m\omega)} \left(\sum_{m=0}^{n} \int_0^{\omega} a(m\omega - s) f(s)\,ds\right) = \sum_{m=1}^{n} \int_0^{\omega} a(t - s + m\omega) f(s)\,ds \tag{2}$$

holds (1) can be re-written as

$$T(t) = T(0)\left(\frac{\sum_{m=0}^{n} a(t + m\omega)}{\sum_{m=0}^{n} a(m\omega)}\right) + \int_0^t a(t - s)f(s)ds \tag{3}$$

Thus we have that (1) has a unique solution whenever (3) does and (2) holds.

Equation (1) can be re-written in the form $T(t) = \int_0^{\omega} K(t,s)f(s)ds$ so for the case where Picard's Theorem is satisfied (see Tricomi [4]) there is a unique solution of minimum norm in $L_2[0,\omega]$ that satisfies the physical constraints of the (n + 1)-tank problem provided that $f(t) \geq 0$ for $0 \leq t \leq \omega$ and $m_0 \geq 0$.

For the case of zero order decay for a(t), $a(t) = E - kt$, $E > 0$, $k > 0$ and $(n + 1)\omega \leq E/k$ equation (1) can be re-written as $T''(t) = -kf(t) + Ef'(t)$, assuming that T and T' are differentiable.

From the above differential equation we have

$$f(t) = g(t) + c\, e^{\frac{k}{E}t}$$

where from equation (1)

$$\begin{aligned} c = [&2(n + 1)\, ET'(0) + n(k\omega - 2E)\, T'(\omega) \\ &+ 2k(n + 1)T(0) - 2knT(\omega) - nE(\omega k - 2E)g(\omega)] \end{aligned}$$

and

$$g(t) = e^{\frac{k}{E}t} \int_0^t \frac{1}{E}\, e^{-\frac{k}{E}s}\, T''(s)ds$$

Uniqueness of the above solution follows from (1) for the above case when $T(t) \equiv 0$; we have $c = 0$. The above will satisfy the (n + 1)-tank problem provided that $f(t) \geq 0$ for $0 \leq t \leq \omega$ and $m_0 \geq 0$.

For the case where $T(t) = \alpha$, α a constant, we have

$$f(t) = 2k\,\alpha\, e^{\frac{k}{E}t} / E\, [2E(n + 1) + n e^{\frac{k}{E}\omega}\, (k\omega - 2E)]$$

and

$$m_0 = \frac{2\alpha + nE\omega(n - 1)[f(\omega) - f(0)] - 2nE\omega f(0)}{(n + 1)\,(2E - n\omega k)}$$

4. GENERAL PROBLEM

Determine conditions under which equation (1) will have a unique solution in $L_2[0,\omega]$ such that $a(t)$ is in $L_2[0,n\omega]$, $f(t) \geq 0$ for $0 \leq t \leq \omega$, $\| f \|$ is a minimum and $m_0 \geq 0$.

REFERENCES

1. Davis, H. T., *The Theory of Linear Operators*, The Principia Press, Bloomington, Indiana (1936).

2. Laider, K. J. and P. S. Bunting, *The Chemical Kinetics of Enzyme Action*, Clavendon Press, Oxford (1973).

3. Lee, G. K. and P. S. Reilly, *The Effect of Slow Pore Diffusion on Observed Enzyme Stability*, paper presented at 84th A.I.Ch.E. National Meeting, Atlanta, Georgia (1978).

4. Tricomi, F. G., *Integral Equations*, Interscience Publishers, Inc., New York (1957).

GLOBAL STABILITY IN ECOLOGICAL MODELS WITH CONTINUOUS TIME DELAY

Wilfred M. Post

Environmental Sciences Division
Oak Ridge National Laboratory
Oak Ridge, Tennessee

Curtis C. Travis

Health and Safety Research Division
Oak Ridge National Laboratory
Oak Ridge, Tennessee

1. INTRODUCTION

The mathematical theory of population ecology has largely been developed employing first order ordinary differential equations. The population growth of a species is assumed to respond instantaneously to the densities of its own population and the populations of other species with which it interacts. However, in ecological models it is often more appropriate to allow the rates of change of the variables to depend on the previous histories of the population densities. For example, the voracity of a predator depends not only on whether there is prey immediately available but also on its physiological state, that is, its level of hunger. The level of hunger depends on some function of its feeding history. Equations incorporating this type of effect have been proposed and analyzed by Bownds and Cushing [1], Cushing [2,3], Gomatam and MacDonald [4], MacDonald [5], May [6], Volterra [7], Wörz-Busekros [8], and others.

We will examine the stability properties of a general system of first order integrodifferential equations which describe the dynamics of interacting species populations. A sufficient condition for the global stability of an equilibrium state is derived. This condition is an improvement over the condition derived by Wörz-Busekros [8] for similar equations in that this condition has intuitive biological interpretations and is verifiable in a finite number of arithmetical steps. We then show that this condition is both necessary and sufficient for global asymptotic stability of the equilibrium for communities of mutualistically interacting species. Application of the results to an ecological system is also provided.

2. TRANSFORMATION TO A DIFFERENTIAL SYSTEM

Suppose that the population dynamics of community of n interacting species is governed by the system of integrodifferential equations,

$$\frac{dx_i}{dt} = x_i(t)(r_i + \sum_{j=1}^{n} a_{ij}x_j(t) + \sum_{j=1}^{n} \gamma_{ij}\int_{-\infty}^{t} g_{ij}(t - s)x_j(s)ds) \tag{1}$$

$$i = 1, 2, \ldots, n$$

where x_i is the population density of the ith species, $r = col(r_1, r_2, \ldots, r_n)$ is the vector of intrinsic population growth rates, $A = (a_{ij})$ and $\Gamma = (\gamma_{ij})$ are interaction matrices, and g_{ij} are memory functions describing the present response of species per capita growth rates to historical population densities. For simplicity it is assumed that every memory function $g_{ij}(t)$ has the form

$$g_{ij}(t) = g_a^k(t) = \frac{a^k}{(k - 1)!} t^{k-1}e^{-at} \tag{2}$$

where a is a positive real parameter and $k \geq 1$ is a fixed positive integer. Under assumption (2) on the form of the memory functions $g_{ij}(t)$, it is possible to change the system of integrodifferential equations (1) into a system of ordinary differential equations [9]. For example, consider a system of the form

$$\frac{dx_i}{dt} = x_i(t)(r_i + \sum_{j=1}^{n} a_{ij}x_j(t)), \qquad i = 1, 2, \ldots, n, \qquad i \neq p \tag{3}$$

$$\frac{dx_p}{dt} = x_p(t)(r_p + \sum_{j=1}^{n} a_{pj}x_j(t) + \gamma_{pm}\int_{-\infty}^{t} x_m(s)g_a^k(t - s)ds) \tag{4}$$

where $1 \leq p,\ m \leq n$. Renaming the lag term as

$$x_{n+k} = \int_{-\infty}^{t} x_m(s) g_a^k(t - s)\,ds \tag{5}$$

we can rewrite the system of equations (3) and (4) by defining k new variables,

$$x_{n+j} = \int_{-\infty}^{t} x_m(s) g_a^j(t - s)\,ds, \qquad j = 1, 2, \ldots, k \tag{6}$$

and differentiating. The additional k equations are linear, that is,

$$\begin{aligned} \frac{dx_{n+1}}{dt} &= a(x_m - x_{n+1}) \\ \frac{dx_{n+j}}{dt} &= a(x_{n+j-1} - x_{n+j}), \qquad j = 2, 3, \ldots, k \end{aligned} \tag{7}$$

This expansion process can obviously be carried out for any number of delay terms in any number of equations with the addition of a suitable number of linear equations. For simplicity of notation, however, we will restrict further analysis to equations with $k = 1$. Thus we will have a system with, at most, 2n equations.

3. STABILITY ANALYSIS

For the case $k = 1$, we obtain a system of the form

$$\begin{aligned} \frac{dx_i}{dt} &= x_i\Big(r_i + \sum_{j=1}^{n} a_{ij}x_j + \sum_{j=1}^{n} \gamma_{ij}x_{j+n}\Big) \qquad i = 1, 2, \ldots, n \\ \frac{dx_i}{dt} &= ax_{i-n} - ax_i \qquad i = n+1, n+2, \ldots, 2n \end{aligned} \tag{8}$$

Let

$$B = \begin{bmatrix} B^{11} & B^{12} \\ B^{21} & B^{22} \end{bmatrix} \tag{9}$$

where $b_{ij}^{12} = -|\gamma_{ij}|$, $B^{21} = \text{diag}(-a)$, $B^{22} = \text{diag}(a)$ and

$$b_{ij}^{11} = \begin{cases} |a_{ii}| & \text{if } i = j \\ -|a_{ij}| & \text{if } i \neq j \end{cases}$$

Before stating the theorem concerning the stability of (1) we present the definition of an M-matrix (see Plemmons [10]).

DEFINITION. A matrix B is an M-matrix if $b_{ij} \leq 0$ for all $i \neq j$ and if any one of the following equivalent statements is true:

(i) All principal minors of B are positive;

(ii) All eigenvalues of B have positive real parts;

(iii) B is nonsingular and $B^{-1} > 0$;

(iv) There exists a vector $u > 0$ such that $Bu > 0$;

(v) There is a vector $v > 0$ such that $B^T v > 0$;

(vi) There exists a positive diagonal matrix D such that $DB + B^T D$ is positive definite.

Notice that the sign pattern of matrix B is that required for an M-matrix. Thus we may use the theory of M-matrices to determine various properties of the system of Eq. (8).

THEOREM 1. Assume that $a_{ii} < 0$ for $i = 1,2,\ldots,n$, then the integrodifferential equation (1) is globally stable in the feasible region if B is an M-matrix.

Proof. Let $\bar{x} = (\bar{x}_1, \bar{x}_2, \ldots, \bar{x}_n)$ be a feasible equilibrium for Eq. (1). Define $\bar{x}_{i+n} = \bar{x}_i$ for $i = 1, 2, \ldots, n$ and let $x_i = \bar{x}_i + y_i$ for $i = 1, 2, \ldots, 2n$. Then

$$\frac{dy_i}{dt} = a_{ii}y_i(\bar{x}_i + y_i) + \sum_{\substack{j=1 \\ j\neq 1}}^{n} a_{ii}(\bar{x}_i + y_i)y_j$$

$$+ \sum_{j=1}^{n} \gamma_{ij}(\bar{x}_i + y_i)y_{j+n} \quad \text{for} \quad i = 1, 2, \ldots, n \tag{10}$$

$$\frac{dy_i}{dt} = ay_{i-n} - ay_i \quad \text{for} \quad i = n+1, n+2, \ldots, 2n$$

Divide this system into two parts:

$$\frac{dy_i}{dt} = f_i(y_i) + g_i(y_1, y_2, \ldots, y_{2n}), \qquad i = 1, 2, \ldots, 2n \tag{11}$$

where

$$f_i(y_i) = \begin{cases} a_{ii} y_i(\bar{x}_i + y_i) & \text{for} \quad i = 1, 2, \ldots, n \\ -ay_i & \text{for} \quad i = n+1, n+2, \ldots, 2n \end{cases} \tag{12}$$

and

$$g_i(y_1, y_2, \ldots, y_{2n}) = \begin{cases} \sum_{\substack{j=1 \\ j \neq i}}^{n} a_{ij}(x_i + y_i)y_j + \sum_{j=1}^{n} \gamma_{ij}(\bar{x}_i + y_i)y_{j+n} \\ \qquad i = 1, 2, \ldots, n \\ ay_{i-n}, \qquad i = n+1, n+2, \ldots, 2n \end{cases}$$

The origin where $(y_1, y_2, \ldots, y_n) = (0, 0, \ldots, 0)$ is a globally stable equilibrium of the decoupled subsystems,

$$\frac{dy_i}{dt} = f_i(y_i), \qquad i = 1, 2, \ldots, 2n$$

in the region $-x_i < y_i$, $i = 1, 2, \ldots, n$. The functions

$$V_i(y_i) = \begin{cases} y_i - \bar{x}_i \ln\left(\dfrac{y_i + \bar{x}_i}{\bar{x}_i}\right), & 1 \leq i \leq n \\ \dfrac{(y_i + \bar{x}_i)^2}{2}, & n+1 \leq i \leq 2n \end{cases} \tag{13}$$

are Lyupanov functions for the subsystems in this region. We prove that under conditions stated in the theorem

$$V(y) = \sum_{i=1}^{n} d_i\left(y_i - x_i \ln\left(\frac{y_i + \bar{x}_i}{\bar{x}_i}\right)\right) + \sum_{i=n+1}^{2n} d_i y_i^2/2 \tag{14}$$

is a Lyupanov function for the system (1) for some choice of positive constants d_i.

$$\frac{d}{dt} V(y(t)) = \sum_{i=1}^{n} d_i \dot{y}_i \left(\frac{y_i}{y_i + \bar{x}_i}\right) + \sum_{i=n+1}^{2n} d_i \dot{y}_i y_i$$

$$= \sum_{i=1}^{n} d_i (a_{ii} y_i^2 + \sum_{\substack{j=1 \\ j\neq 1}}^{n} a_{ij} y_i y_j + \sum_{j=1}^{n} \gamma_{ij} y_i y_{j+n}$$

$$+ \sum_{i=n+1}^{2n} d_i (a y_i y_{i-n} - a y_i^2)$$

$$\leq \sum_{i=1}^{n} d_i (a_{ii} y_i + \sum_{\substack{j=1 \\ j\neq 1}}^{n} |a_{ij}|\,|y_i|\,|y_j| + \sum_{j=1}^{n} |\gamma_{ij}|\,|y_i|\,|y_{j+n}|)$$

$$+ \sum_{i=n+1}^{2n} d_i (-a y_i^2 + a|y_i|\,|y_{i-n}|)$$

$$= -\sum_{i,j=1}^{2n} |y_i| C_{ij} |y_j|$$

where C_{ij} is the i, jth element of the symmetric matrix $DB + B^T D$. Since, by assumption, B is an M-matrix, condition (vi) implies that there exists a choice of positive constants d_i such that $DB + B^T D$ is positive definite, where $D = \text{diag}(d_i)$.

4. MUTUALISTIC SYSTEMS

Local stability of system (1) is determined by the eigenvalues of the Jacobian matrix of (8) evaluated at the equilibrium $\bar{x}$,

$$J = \text{diag}(\bar{x}_1, \bar{x}_2, \ldots, \bar{x}_n, 1, 1, \ldots, 1)A$$

where

$$A = \begin{bmatrix} A^{11} & A^{12} \\ A^{21} & A^{22} \end{bmatrix} \tag{15}$$

and $a_{ij}^{11} = a_{ij}$, $a_{ij}^{12} = \gamma_{ij}$, $A^{21} = \text{diag}(a)$, $A^{22} = \text{diag}(-a)$. Previously the authors have established necessary and sufficient conditions for the local stability of mutualistic communities [11]. A mutualistic community is one in which interactions between species are nonnegative. We will say that Eq. (1) represents a mutualistic community if $a_{ij} \geq 0$, $i \neq j$, and the delay terms of the species interactions are also mutualistic, that is, $\gamma_{ij} \geq 0$. The following theorem can be established as in Travis and Post [11].

THEOREM 2. The feasible equilibrium of a mutualistic community represented by the integrodifferential equation (1) is locally stable if and only if -A (A given by (15)) is an M-matrix.

Suppose that the integrodifferential equation (1) represents a mutualistic community. By Theorem 2 a feasible equilibrium of this system is locally stable if and only if -A is an M-matrix. However, from Theorem 1, the feasible equilibrium of a mutualistic community is globally stable if -A is an M-matrix. Since local and global stability of a mutualistic system depend on the same conditions, we have established that local stability implies global stability. Since the converse is always true, we have established the validity of the following corollary to Theorem 1.

COROLLARY 1. A feasible equilibrium of a mutualistic community described by Eq. (1) is globally stable in the feasible region if and only if it is locally stable.

5. EXAMPLE

The following equations describe a hypothetical predator-prey interaction.

$$\begin{aligned} \frac{dx_1}{dt} &= x_1(t)(r_1 - a_{11}x_1(t) - a_{12}x_2(t)) \\ \frac{dx_2}{dt} &= x_2(t)(-r_2 - a_{22}x_2(t) + \gamma_{21}\int_{-\infty}^{t} g_a^1(t - s)x_1(s)ds) \end{aligned} \tag{16}$$

The growth rate of the prey (x_1) is instantaneously reduced through interaction with the predator. The growth rate of the predator (x_2), on the other hand, is increased by its previous feeding history. This introduces a continuous delay term into the Lotka-Volterra equations.

The equilibrium

$$\bar{x} = (\bar{x}_1, \bar{x}_2)^T = \left(\frac{r_1a_{22} + r_2a_{12}}{a_{11}a_{22} + a_{12}\gamma_{21}}, \quad \frac{r_1\gamma_{21} - r_2a_{11}}{a_{11}a_{22} + a_{12}\gamma_{21}}\right)^T \tag{17}$$

is feasible if $r_1\gamma_{21} > r_2a_{11}$. With $k = 1$, the above equations are globally asymptotically stable if $0 < a_{21}\gamma_{21}/a_{11}a_{22} < 1$. This is easily computed using property (i) of an M-matrix where

$$B = \begin{bmatrix} a_{11} & -|-a_{12}| & 0 \\ 0 & a_{22} & -|\gamma_{21}| \\ -a & 0 & a \end{bmatrix}$$

6. CONCLUSIONS

The sufficient conditions for global stability of the equilibrium of ecological equations presented here are restrictive. Wörz-Busekros [8] derived a similar condition for Eq. (16) and found the region of stability to be at least $0 < a_{12}\gamma_{21}/a_{11}a_{22} < 8$. This is considerably larger than the region of parameter space described using Theorem 1. In Post and Travis [12], the authors point out that real ecological systems are not randomly constructed. They have long evolutionary histories during which destabilizing interactions have been eliminated. Thus, real systems are characterized by parameters that occupy the relatively small and sparse regions of parameter space possessing enhanced stability properties. This region is usually much smaller than that theoretically predicted on the basis of stability considerations alone. It is shown that a large proportion of ecologically plausible models which are stable satisfy the conditions in Theorem 1. Thus, we conjecture that most of the parameters in the region $1 < a_{12}\gamma_{21}/a_{11}a_{22} < 8$ give rise to models which, while globally stable, are not generally ecologically plausible.

The ecological implications of Theorem 1 are summarized by property (iv) in the definition of an M-matrix. This condition implies that the system (1) is stable if there exists positive constants $d_1, d_2, \ldots, d_n$ such that

$$-d_i a_{ii} > \sum_{\substack{j=1 \\ j \neq i}}^{n} d_j |a_{ij}| + \sum_{k=1}^{n} d_k |\gamma_{ik}|, \qquad i = 1, 2, \ldots, n$$

From this we conclude that an ecological community will be globally stable if intraspecific regulation of population densities is stronger than some positive linear combination of the effect of interspecific interactions.

While Theorem 1 requires strong restrictions on the equation parameters, weaker sufficient conditions are not known that are interpretable in ecological terms, verifiable in a finite number of arithmetical steps, and applicable to all possible patterns of interspecies interactions.

ACKNOWLEDGMENTS

Publication No. 1409, Environmental Sciences Division, Oak Ridge National Laboratory.

Research supported by the National Science Foundation's Ecosystem Studies Program under Interagency Agreement No. DEB 77-25781 with the U.S. Department of Energy under contract W-7405-eng-26 with Union Carbide Corporation.

REFERENCES

1. Bownds, J. M., and J. M. Cushing. 1975. On the behavior of solutions of predator-prey equations with hereditary terms. *Math. Biosci.* 25: 41-54.

2. Cushing, J. M. 1976. Periodic solutions of two species interaction models with lags. *Math. Biosci.* 31:143-156.

3. Cushing, J. M. 1976. Predator prey interactions with time delays. *J. Math. Biol.* 3:359-380.

4. Gomatam, J., and N. MacDonald. 1975. Time delays and the stability of two competing species. *Math. Biosci.* 24:247-255.

5. MacDonald, N. 1978. Time Lags in Biological Models. Lecture Notes in Biomathematics No. 27, Springer-Verlag, Berlin.

6. May, R. M. 1973. Time delay versus stability in population models with two and three trophic levels. *Ecology* 54:315-325.

7. Volterra, V. 1931. *Lecon sur la théorie mathématique de la lutte pour la vie.* Gauthier-Villars, Paris.

8. Wörz-Busekros, A. 1978. Global stability in ecological systems with continuous time delay. *SIAM J. Appl. Math.* 35:123-134.

9. Fargue, M. D. 1973. Réducibilité des systèmes héréditaires à des systèmes dynamiques. *C. R. Acad. Sci. Paris Ser. B* 277:471-473.

10. Plemmons, R. V. 1977. M-matrix characterizations. I. Nonsingular M-matrices. *Lin. Alg. and Appl.* 18:175-188.

11. Travis, C. C., and W. M. Post. 1979. Dynamics and comparative statics of mutualistic communities. *J. Theor. Biol.* 78:553-571.

12. Post, W. M., and C. C. Travis. 1979. Quantitative stability in models of ecological communities. *J. Theor. Biol.* 79:547-553.

POLYNOMIAL APPROXIMATIONS TO EXPONENTIAL STABILITY EQUATIONS

William Squire

Department of Aerospace Engineering
West Virginia University
Morgantown, West Virginia

1. INTRODUCTION

Many systems of engineering importance have components which do not respond instantaneously to changes in the operating conditions [1]. When such lags are taken into account the systems are described by functional differential equations (also called differential-difference equations or differential equations with retarded arguments) instead of by ordinary differential equations. When dealing with a time-invariant system, the ordinary differential equation has a polynomial characteristic equation. The methods for analyzing the stability of such equations dating back to Routh [2] and Hurwitz [3] are well known. The corresponding functional differential equations have transcendental characteristic equations involving exponential terms as well as powers. Such exponential polynomials have an infinite number of roots so the stability problem is more difficult.

A simple approach, dating back to the early work of Minorsky [4], expands the exponentials in Taylor series and truncates the expansion. The techniques for analyzing polynomials are then applied. The dangers of this approach were soon recognized [5] and exact methods for dealing directly with exponential polynomials were developed [6-8]. However these methods are graphical or numerical and are applied to expressions with specific numerical values of the parameters. The engineer would prefer to express the stability boundary as general relations between the parameters of the problem. For this reason the Taylor series approach is still used in engineering work. Recently Mutharasan and Luus [9] have developed a modification which improved the accuracy appreciably in some cases.

This paper is an extension of an investigation by Chosky [10] who compared the results of the Taylor series expansion applied to the mathematically equivalent equations:

$$S + Ke^{-S} = 0 \tag{1a}$$

and

$$Se^{S} + K = 0 \tag{1b}$$

2. ANALYTICAL BACKGROUND

An alternative approach is to consider different approximations to the exponential rather than different forms of the equation. We will use:

$$e^{-x} = 1 - x + x^2/2 - x^3/6 + \ldots \tag{2a}$$

which will be referred to as the Taylor expansion,

$$e^{-x} = (1 + x + x^2/2 + x^3/6 + \ldots)^{-1} \tag{2b}$$

and the rational function form

$$e^{-x} = \frac{1 - x/2 + x^2/8 - x^3/48 + \ldots}{1 + x/2 + x^2/8 + x^3/48 + \ldots} \tag{2c}$$

It is easily seen that the combination of (1a) with (2b) is equivalent to using (1b) and (2a). Using (2c) with either (1a) or (1b) is equivalent to the use of the direct Taylor expansion with the symmetric form

$$Se^{S/2} + Ke^{-S/2} = 0 \tag{3}$$

The stability of a polynomial can be determined without solving for the roots by using the Hurwitz criteria. The condensed form due to Lienard and Chipart [11] states: The polynomial

$$a_0 + a_1 z + a_2 z^2 + a_3 z^3 + \ldots$$

is stable if

(a) all the coefficients are positive

(b) $a_1 a_2 - a_0 a_3 > 0$ for a cubic

$a_3(a_1 a_2 - a_0 a_3) - a_1^2 a_4 > 0$ for a quartic

both $(a_1 a_2 - a_0 a_3)$ and $(a_1 a_2 - a_0 a_3)(a_2 a_5 - a_3 a_4)$
$- a_1 a_4 (a_1 a_4 - a_0 a_5) > 0$ for a quintic.

The criteria can be extended to higher order polynomials by the Hurwitz determinant but become increasing complex.

3. SPECIFIC EXAMPLES

The exact stability criterion for Eq. (1) is that $0 \leq K \leq \pi/2 = 1.5708$. Using (1a) and (2a) gives the polynomial

$$K + (1 - K)S + KS^2/2 - KS^3/6 + KS^4/24 + \ldots \tag{4}$$

Truncating to a linear or quadratic equation gives $0 \leq K \leq 1$ as the stability criterion. For all higher order approximations this approach predicts instability for all K. This is the type of result that led to distrust of the Taylor series approach.

The Mutharasan and Luus method eliminates the S^3 term between the cubic and S times the quadratic. This gives

$$K + [(1 - K) + K/3]S + [K/2 + (1 - K)/3]S^2 = 0 \tag{5}$$

The requirement that the coefficient of the linear term be positive gives the condition $K \leq 1.5$ which is a considerable improvement.

Using Eq. (2b) gives the polynomial approximation

$$0 = K + S + S^2 + \frac{S^3}{2} + \frac{S^4}{6} + \frac{S^5}{24} + \dots \qquad (6)$$

The linear and quadratic equations indicate stability for all K. The higher order approximations give:

order	3	4	5	6	7 and higher
limiting K	2.0	4/3	1.464	1.6	unstable

Using Eq. (2c) gives

$$0 = K + S(1 - \frac{K}{2}) + S^2(\frac{1}{2} + \frac{K}{8}) + S^3(\frac{1}{8} - \frac{K}{48}) + \dots \qquad (7)$$

This gives the sequence 2, 1.464, 1.5331 which is quite accurate. However because K is now in all the coefficients the determination of the numerical value of K for the higher order approximations involves solving a high order polynomial numerically. It was not possible to determine if for some high order approximation this procedure would also break down.

Some improvement was obtained by using

$$e^{-x} = \frac{1 - x/2 + x^2/12}{1 + x/2 + x^2/12} \qquad (8)$$

This is a member of a family of rational approximations given by Lanczos [12]. Substituting into (1a) gives the cubic

$$S^3/12 + (1/2 + K/12)S^2 + (1 - K/2)S + K = 0 \qquad (9)$$

The Hurwitz criteria reduces to

$$K^2 + 6K + 12 = 0 \qquad (10)$$

The solution K = 1.583 is a significant improvement over the 1.464 obtained from Eq. (2c) with the same number of terms, which is entirely due to changing the third term from $x^2/8$ to $x^2/12$.

The second example

$$S^2 + 2S + 1 + Ke^{-aS} = 0 \tag{11}$$

is taken from Mutharasan and Luus [9]. We now need not a numerical value for K but a relation between K and a that will define the stability boundary. Using either $e^{-aS} = 1 - aS$ or $1 - aS + a^2S^2/2$ gives the condition $K = 2/a$. Using more terms gives instability because of the negative coefficient of the S^3 term.

Eliminating the S^3 term between the cubic and S times the quadratic gives the quadratic equation

$$\begin{aligned} 0 = 1 + K + [2 - aK + \frac{Ka^3/6(1 + K)}{1 + Ka^2/2}]S \\ + [1 + Ka^2 + \frac{(2 - ak)Ka^3/6}{1 + Ka^2/2}]S^2 \end{aligned} \tag{12}$$

The linear term gives the stability condition

$$a(Ka)^2/3 + (1 - a - a^2/6)Ka - 2 = 0 \tag{13}$$

or

$$K = \frac{2}{a}\,\frac{2}{[(1 - a - a^2/6)^2 + 8/3a]^{1/2} + 1 - a - a^2/6} \tag{14}$$

Using $e^{-aS} = (1 - aS/2)/(1 + aS/2)$ gives

$$aS^3/2 + (1 + a)S^2 + [2 + a(1 - K)/2]S + 1 + K = 0 \tag{15}$$

The Hurwitz criterion is

$$[2 + a(1 - K)/2]\ (1 + a) - (1 + K)a/2 = 0 \tag{16}$$

a linear equation in K which gives

$$K = \frac{2}{a} \frac{1 + a - a^2/2}{1 + a/2} \tag{17}$$

Table 1 compares this result, Eq. (14) and the simple 2/a expression with the exact value obtained by a graphical procedure. The accuracy of (14) is very good and (17) based on the rational approximation is again better than the Taylor series approach.

TABLE 1
Comparison of K Values

a	Exact	2/a	Eq. (17)	Eq. (14)
.1	20.7282	20.0	20.8571	20.6770
.2	10.6957	10	10.7273	10.6859
.3	7.3754	6.6667	7.2754	7.3599
.4	5.6911	5.0	5.5	5.6993
.5	4.6931	4.0	4.4	4.7040
.6	4.0292	3.3333	3.5410	4.0408
.7	3.5561	2.8751	3.0794	3.5671

Using Eq. (8) would also give a quadratic for K and a result similar to but somewhat more complicated than Eq. (14). The use of (2b) leads to even more complicated algebra.

4. DISCUSSION

It appears that both the use of the rational function approximation and the Mutharasan and Luus modification give a significant improvement over the simple Taylor series approach. The second example clearly shows why engineers will favor the approximate approach over the exact computational approach, particularly for preliminary design.

It would be desirable for the mathematicians to produce some general theorems which would enable the engineer to know when a general prediction such as the complete stability obtained from Eq. (6) when truncated to a linear or quadratic equation and the instability when seven or more terms were kept are invalid. Given safeguards against such misleading results the engineer could usually sacrifice some accuracy for ease of calculation and the generality of an analytical expression.

REFERENCES

1. F. Marble and D. Cox, Servo-Stabilization of Low-Frequency Oscillations in a Liquid Bipropellant Rocket Motor, *J. Am. Rocket Soc.*, 23 (1953) pp. 63-74 and 81.

2. E. J. Routh, *On the Stability of a Given State of Motion*, Macmillan, Londen, 1877.

3. A. Hurwitz, Ueber die Bedingungen unter welchen eine Gleichung nur Wurzeln mit negativen reellen Theilen besitzt, *Math. Ann.*, 46 (1895) pp. 273-284.

4. N. Minorsky, Control Problems, *J. Franklin Inst.*, 232 (1941) pp. 519-551.

5. S. Sherman, A note on stability calculation and time lag, *Quart. Appl. Math.*, 5 (1941) pp. 92-97.

6. L. S. Pontryagin, On Zeroes of Some Transcendental Functions, *Izv. Akad. Nauk. SSSR.*, 6 (1942) pp. 115-134.

7. M. Satche, Discussion of Stability of Linear Oscillating Systems, *J. Appl. Mech.*, 16 (1949) pp. 419-420.

8. B. Porter, *Stability Criteria for Linear Systems*, Academic Press, New York, 1968.

9. R. Mutharasan and R. Luus, Analysis of Time-Delay Systems by Series Approximations, *AICHE Journal*, 21 (1975) pp. 567-571.

10. H. H. Chosky, *Time Lag Systems* in Progress in Control Engineering, R. H. McMillan editor, Heywood, 1962.

11. A. Lienard and M. H. Chipart, On the Sign of the Real Part of the Roots of Algebraic Equations, *J. Math. Pure Appl.*, 6 (1914) pp. 291-346.

12. C. Lanczos, *Applied Analysis*, Prentice-Hall, Englewood Cliffs, 1956.

REGULARITY OF SOLUTIONS TO AN INHOMOGENEOUS DIFFERENTIAL EQUATION IN BANACH SPACE

Curtis C. Travis

Technology Assessments Section
Health and Safety Research Division
Oak Ridge National Laboratory
Oak Ridge, Tennessee

1. INTRODUCTION

Let $T(t)$, $t \geq 0$, be a strongly continuous semigroup of bounded linear operators in the Banach space X with infinitesimal generator A and let f be a continuous X-valued function on $(0,\infty)$. It is well-known that without some restrictions on the semigroup $T(t)$, $t \geq 0$, or the continuous function f, the weak solution

$$v(t) = T(t)x + \int_0^t T(t - s)f(s)ds \tag{1.1}$$

need not be a strong solution of the inhomogeneous linear differential equation

$$\begin{aligned} du(t)/dt &= Au(t) + f(t) \\ u(0) &= x \end{aligned} \tag{1.2}$$

R. Phillips [6] has shown that if $x \in D(A)$ and f is continuously differentiable, then the weak solution (1.1) is the unique strong solution of (1.2). The condition that f be continuously differentiable can be weakened by requiring further conditions on the strongly continuous semigroup

$T(t)$, $t \geq 0$. For example, it is known [4] that (1.1) is a strong solution of (1.2) for every Hölder continuous function f if $T(t)$, $t \geq 0$, is holomorphic. Webb [8] has established that $T(t)\,X \subset D(A)$ for $t > 0$ is a sufficient condition for a weak solution to be a strong solution for every continuous function of bounded variation. Beirao Da Veiga has recently shown [1] that in a reflexive Banach space X, the condition $T(t)X \subset D(A)$ for $t > 0$ is not needed. That is, if $x \in D(A)$ and f is a continuous function of bounded variation, then the weak solution (1.1) is a strong solution of (1.2).

Our objective in this paper is to characterize the class of strongly continuous semigroups for which a weak solution of equation (1.2) is a strong solution when $f \in C([0,r]; X)$. We demonstrate that (1.1) is a strong solution of (1.1) for every continuous function f if and only if the semigroup $T(t)$, $t \geq 0$, is of bounded semivariation.

2. OPERATORS OF BOUNDED SEMIVARIATION

Given a closed interval [a,b] of the real line, a subdivision of [a,b] is a finite sequence $d : a = d_0 < d_1 < \ldots < d_n = b$. Let D[a,b] denote the set of all subdivisions of [a,b], and X and Y be Banach spaces. For $\alpha : [a,b] \to L(X;Y)$ and $d \in D[a,b]$, define

$$SV_d[\alpha] = \sup\{\|\sum_{i=1}^{n} [\alpha(d_i) - \alpha(d_{i-1})]x_i\| \mid x_i \in X, \|x_i\| \leq 1\}$$

and

$$SV[\alpha] = \sup\{SV_d[\alpha] \mid d \in D[a,b]\}$$

We say α is of bounded semivariation (see [3]) if $SV[\alpha] < \infty$.

It follows from

$$SV_d[\alpha] = \sup\{\|\sum_{i=1}^{n} [\alpha(d_i) - \alpha(d_{i-1})]x_i\| \mid x_i \in X, \|x_i\| \leq 1\}$$

$$\leq \sum_{i=1}^{n} \|\alpha(d_i) - \alpha(d_{i-1})\|$$

that the concept of bounded semivariation is an extension of the concept of bounded variation. For functions, the concepts of bounded semivariation and bounded variation are equivalent. In fact, if $f : [a,b] \to X \subset L(X';C)$ and $\langle\cdot,\cdot\rangle$ is the pairing between X and X', then

$$SV_d[f] = \sup\{|\sum_{i=1}^{n} \langle f(d_i) - f(d_{i-1}), x_i\rangle| \ |x_i \in X', \|x_i\| \le 1\}$$

$$= \sum_{i=1}^{n} \| f(d_i) - f(d_{i-1}) \|$$

The following proposition is stated without proof.

PROPOSITION 2.1 (Hönig): If $\alpha : [a,b] \to L(X;Y)$ is of bounded semivariation and $f : [a,b] \to X$ is continuous, then the Riemann-Stieltjes integral $\int_a^b d\alpha(s)f(s)$ exists.

LEMMA 2.1. If $\alpha : [0,r] \to L(X;Y)$ is of bounded semivariation, $\alpha(t)$ is strongly continuous at $t = 0$, and $f : [0,r] \to X$ is continuous, then

$$\lim_{t\to 0^+} \int_0^t d\alpha(s)\, f(s) = 0$$

3. REGULARITY OF SOLUTIONS

When F is the class of Hölder continuous functions, Crandall and Pazy [2] demonstrate that

(i) $T(t)X \subset D(A)$ for $t > 0$

(ii) $\lim_{t\to 0^+} t^p \| AT(t) \| < \infty$ for all $p > 1$

are necessary and sufficient conditions on the semigroup T(t) to assure that the function (1.1) is a strong solution of the Cauchy problem (1.2) for every $f \in F$. Our principal result will be to establish bounded semivariation as a necessary and sufficient condition on the semigroup T(t), $t \ge 0$, to assure that (1.1) is a strong solution of (1.2) when F is the class of continuous functions. We first establish some necessary lemmas.

LEMMA 3.1. If f is continuous and $T(\cdot)$ is of bounded semivariation on $[0,r]$, then for $t \in [0,r]$,

$$\int_0^t T(t - s)f(s)ds \in D(A) \tag{3.1}$$

and

$$A \int_0^t T(t - s)f(s)ds = \int_0^t d_sT(t - s)f(s) \tag{3.2}$$

Proof. Under the assumptions on f and $T(\cdot)$, it follows from Proposition 2.1 that the Riemann-Stieltjes integral $\int_0^t d_sT(t - s)f(s)$ exists. For a fixed positive integer n, let $d_i^n = it/n$, where $i = 0, 1, \ldots, n$. Define $g_n(s) = T(t - s)f(d_i^n)$ for $d_{i-1}^n < s \leq d_i^n$ and $g_n(0) = T(t)f(0)$. Since the sequence $\{g_n\}$ is uniformly bounded and converges to $T(t - s)f(s)$ on $[0,t]$,

$$\lim_{n\to\infty} \int_0^t g_n(s)ds = \int_0^t T(t - s)f(s)ds \tag{3.3}$$

Since $\int_0^t g_n(s)ds \in D(A)$ and

$$\begin{aligned} A \int_0^t g_n(s)ds &= \sum_{i=1}^{n} [T(t - d_i^n) - T(t - d_{i-1}^n)]f(d_i^n) \\ &\to \int_0^t d_sT(t - s)f(s) \end{aligned} \tag{3.4}$$

(3.1) and (3.2) follow from the closedness of A.

LEMMA 3.2. If f is continuous and $T(\cdot)$ is of bounded semivariation on $[0,r]$, then $\int_0^t d_sT(t - s)f(s)$ is continuous in t on $[0,r]$.

Proof. Assume $0 \leq t < r$ and $0 < \Delta t$, and observe that

$$\begin{aligned} &\int_0^{t+\Delta t} d_sT(t + \Delta t - s)f(s) - \int_0^t d_sT(t - s)f(s) \\ &\quad = [T(\Delta t) - I] \int_0^t d_sT(t - s)f(s) \\ &\quad + \int_0^{\Delta t} d_sT(t - s)f(t + \Delta t - s) \end{aligned}$$

Since it follows from Lemma 2.1 that

$$\lim_{\Delta t \to 0^+} \int_0^{\Delta t} d_s T(t - s) f(t + \Delta t - s) = 0$$

continuity of $\int_0^t d_s T(t - s) f(s)$ from the right is a consequence of the fact that $\lim_{h \to 0^+} [T(h) - I]x = 0$ for all $x \in X$.

To show continuity from the left, let $0 < t \leq r$ and $0 < \Delta t$. Then

$$\begin{aligned}
&\int_0^t d_s T(t - s) f(s) - \int_0^{t-\Delta t} d_s T(t - \Delta t - s) f(s) \\
&\quad = \int_0^{t-\Delta t} d_s T(t - \Delta t - s)(T(\Delta t) - I) f(s) + \int_{t-\Delta t}^t d_s T(t - s) f(s) \\
&\quad = \int_0^{t-\Delta t} d_s T(s)(T(\Delta t) - I) f(t - \Delta t - s) + \int_0^{\Delta t} d_s T(s) f(t - s)
\end{aligned}$$

Since

$$\begin{aligned}
&\left\| \int_0^{t-\Delta t} d_s T(s)(T(\Delta t) - I) f(t - \Delta t - s) \right\| \\
&\quad \leq SV[T(\cdot)] \sup_{0 \leq s \leq t-\Delta t} \| (T(\Delta t) - I) f(s) \|
\end{aligned}$$

and

$$\lim_{\Delta t \to 0^+} \int_0^{\Delta t} d_s T(s) f(t - s) = 0$$

continuity from the left follows from the uniform continuity of f on $[0,r]$ and the fact that $\lim_{h \to 0^+} (T(h) - I)z = 0$ uniformly for z in a compact subset of X.

PROPOSITION 3.1. The function (1.1) is a strong solution of the Cauchy problem (1.2) for every continuous function f if and only if $T(\cdot)$ is of bounded semivariation on $[0,r]$.

Proof. It is easily shown (Pazy [5]) that the weak solution v defined by (1.1) is a strong solution of (1.2) if and only if $v(t) \in D(A)$ and $Av(t)$ is continuous on $[0,r]$. If f is continuous and $T(\cdot)$ of bounded semivariation on $[0,r]$, then these conditions are satisfied by virtue of Lemmas 3.1 and 3.2.

To demonstrate the necessity of the assumption that $T(\cdot)$ be of bounded semivariation on $[0,r]$, assume $SV[T(\cdot)] = \infty$ on $[0,r]$ and define the bounded linear operator $L : C([0,r]; X) \to X$ by $L(f) = \int_0^r T(r - s)f(s)ds$. Since every weak solution is a strong solution, $L(f) \in D(A)$ for every continuous f. It follows from the closedness of A and the boundedness of L that $AL : C([0,r]; X) \to X$ is bounded. Since $SV[T(\cdot)] = \infty$, there exists a sequence $d_n \in D$ such that $SV_{d_n}[T(\cdot)] > N$ and $\Delta d_n \leq 1/N$. Define $f_{d_n}(t) = f(d_i^n)$ if $d_{i-1}^n < t \leq d_i^n$ and define

$$L_n(f) = \int_0^r T(r - s)f_{d_n}(s)ds$$

Since $AL_n(f) \to AL(f)$ for fixed $f \in C([0,r]; X)$, it follows from the Banach-Steinhaus theorem that there exists M such that $\| AL_n \| \leq M$ for all $n \geq 1$. But

$$\| AL_n(f) \| = \| A \int_0^r T(r - s)f_{d_n}(s)ds \|$$

$$= \| \sum_{i=1}^{n} A \int_{d_{i-1}^n}^{d_i^n} T(r - s)f(d_i^n)ds \|$$

$$= \| \sum_{i=1}^{n} [T(r - d_i^n) - T(r - d_{i-1}^n)]f(d_i^n) \|$$

Thus

$$\| AL_n \| = \sup\{\| \sum_{i=1}^{n} [T(r - d_i^n) - T(r - d_{i-1}^n)]f(d_i^n) \| \mid f \in C([0,r]; X); \| f \| \leq 1\}$$

The proof will be complete if we show that for fixed $d = \{s_i\} \in D$,

$$\sup\{\| \sum_{i=1}^{n} [T(r - s_i) - T(r - s_{i-1})]f(s_i) \| \mid f \in C([0,r]; X); \| f \| \leq 1\}$$

$$= \sup\{\| \sum_{i=1}^{n} [T(r - s_i) - T(r - s_{i-1})]x_i \| \mid x_i \in X; \| x_i \| \leq 1\}$$

$$= SV_d[T(\cdot)]$$

That

$$\sup\{\| \sum_{i=1}^{n} [T(t - s_i) - T(t - s_{i-1})]f(s_i) \| \mid f \in C([0,r]; X), \| f \| \leq 1\}$$

$$\leq \sup\{\| \sum_{i=1}^{n} [T(t - s_i) - T(t - s_{i-1})]x_i \| \mid x_i \in X, \| x_i \| \leq 1\}$$

follows easily. To see the converse, for fixed but arbitrary $x_i \in X$, $i = 1, \ldots, n$, $\| x_i \| \leq 1$, define $f(s) = (1 - \lambda)x_{i-1} + \lambda x_i$ for $s = (1 - \lambda)s_{i-1} + \lambda s_i$, $0 \leq \lambda \leq 1$. Then

$$\| \sum_{i=1}^{n} [T(t - s_i) - T(t - s_{i-1})]x_i \|$$

$$= \| \sum_{i=1}^{n} [T(t - s_i) - T(t - s_{i-1})]f(s_i) \|$$

$$\leq \sup\{\| \sum_{i=1}^{n} [T(t - s_i) - T(t - s_{i-1})]f(s_i) \| \mid f \in C([0,r]; X),$$
$$\| f \| \leq 1\}$$

The result now follows.

The characterization of strongly continuous semigroups of bounded semivariation remains an open question. However, we can establish some partial results in this direction.

LEMMA 3.3. A necessary condition for a strongly continuous semigroup $T(\cdot)$ to be of bounded semivariation on $[0,r]$ is that $T(t)X \subset D(A)$ for $t > 0$.

Proof. It suffices to prove that $T(t)X \subset D(A)$ for $t \in (0,r)$ since $T(t_0)X \subset D(A)$ implies $T(t)X \subset D(A)$ for $t \geq t_0$. Suppose there exist $x_0 \in X$ and $t_0 \in (0,r)$ such that $T(t_0)x_0 \notin D(A)$ and consider the Cauchy problem $u'(t) = Au(t) + f(t)$, $u(0) = 0$, where $f(s) = T(s)x_0$. The unique weak solution of this equation given by $u(t) = tT(t)x_0$ is not a strong solution since it is not differentiable at $t = t_0$. This is in contradiction to Proposition 3.1 since $f \in C([0,r]; X)$.

LEMMA 3.4. If a strongly continuous semigroup is of bounded semivariation on $[0,r]$, $r > 0$, then it is of bounded semivariation on $[0,a]$ for any $a > 0$.

Proof. If $a > r$, then it is easily established that

$$SV_{[0,a]}[T(\cdot)] \leq SV_{[0,r]}[T(\cdot)] + SV_{[r,a]}[T(\cdot)]$$

As a consequence of Lemma 3.3, $T(t)$, $T \geq 0$, is continuously differentiable in the uniformly operator topology for $t > 0$, and, consequently, since $r > 0$, there exists $M > 0$ such that $\| T'(s) \| \leq M$ for $s \in [r,a]$. Thus if $\{d_i\}$ is a subdivision of $[r,a]$ and $\| x_i \| \leq 1$, then

$$\| \sum_{i=1}^{n} [T(d_i) - T(d_{i-1})]x_i \|$$

$$= \| \sum_{i=1}^{n} \int_{d_{i-1}}^{d_i} T'(s)x_i ds \| \leq M(a - r)$$

It follows that $SV_{[r,a]}[T(\cdot)] < \infty$, and validity of the lemma is established.

We can thus define a strongly continuous semigroup to be of bounded semivariation if it is of bounded semivariation on some, and hence all, intervals of the form $[0,r]$, $r > 0$. Our next proposition characterizes strongly continuous groups of bounded semivariation.

PROPOSITION 3.2. A strongly continuous group is of bounded semivariation if and only if its infinitesimal generator is bounded.

Proof. Let A denote the infinitesimal generator of a strongly continuous group $T(t)$, $t \in R$, and assume A is bounded. Then for arbitrary $d \in D[-r,r]$,

$$\| \sum_{i=1}^{n} [T(s_i) - T(s_{i-1})]x_i \| = \| \sum_{i=1}^{n} \int_{s_{i-1}}^{s_i} T'(s)x_i ds \|$$

$$= \| \sum_{i=1}^{n} A \int_{s_{i-1}}^{s_i} T(s)x_i ds \| \leq \| A \| M \int_{-r}^{r} e^{\omega s} ds$$

implying that $T(\cdot)$ is of bounded semivariation on $[-r,r]$.

To prove the converse, suppose $T(t)$, $t \in R$, is of bounded semivariation. By Lemma 3.3, $T(t)X \subset D(A)$ for $t > 0$. Thus for $x \in X$ and $t_0 \neq 0$,

$x = T(0)x = T(t_0)T(-t_0)x \in D(A)$, implying that $D(A) = X$. The boundedness of A now follows from the closed graph theorem.

COROLLARY 3.1. Suppose the operator A in equation (1.1) is the infinitesimal generator of a strongly continuous group. Then a weak solution of (1.1) is a strong solution for every continuous function f if and only if A is bounded.

ACKNOWLEDGMENT

The Oak Ridge National Laboratory is operated by Union Carbide Corporation under contract W-7405-eng-26 with the U.S. Department of Energy.

REFERENCES

1. H. Beirao Da Veiga, Inhomogeneous Evolution Equations in Banach Spaces with a Bounded Variation Data, *J. Nonlin. Anal.* (in press).

2. M. G. Crandall and A. Pazy, On the differentiability of weak solutions of a differential equation in Banach space, *J. Math. Mech.* 18 (1969), 1007-1016.

3. C. S. Hönig, The Green function of a linear differential equation with a lateral condition, *Bull. Am. Math. Soc.* 79 (1973), 587-593.

4. T. Kato, *Perturbation theory for linear operators*, Springer-Verlag, New York, 1966.

5. A. Pazy, *Semi-groups of linear operators and applications to partial differential equations*, Lecture Notes #10, University of Maryland (1974).

6. R. S. Phillips, Perturbation theory for semi-group of linear operators, *Trans. Am. Math. Soc.* 74 (1953), 199-221.

7. C. C. Travis, An abstract Volterra-Stieltjes integral equation, *Volterra Equations*, Lecture Notes in Mathematics, Vol. 737, (1980), pp. 287-294, Springer-Verlag, New York.

8. G. F. Webb, Regularity of solutions to an abstract inhomogeneous linear differential equation, *Proc. Am. Math. Soc.* 62 (1977), 271-277.

ON THE STRUCTURE OF RESOLVENTS OF VOLTERRA EQUATIONS

Robert L. Wheeler*

Department of Mathematics
University of Missouri
Columbia, Missouri

1. INTRODUCTION

The integral resolvent $r_1(t)$ and differential resolvent $r_2(t)$ associated with the linear Volterra integral and integrodifferential systems

$$x(t) = f(t) - \int_0^t x(t - s)B(s)ds, \qquad t \in R^+ \equiv [0,\infty) \tag{1.1}$$

$$x'(t) = x(t)A + \int_0^t x(t - s)B(s)ds, \qquad x(0) = x_0, \qquad t \in R^+ \tag{1.2}$$

are defined as the solutions of the equations

$$r_1(t) = B(t) - \int_0^t r_1(t - s)B(s)ds, \qquad t \in R^+ \tag{1.3}$$

$$r_2'(t) = r_2(t)A + \int_0^t r_2(t - s)B(s)ds, \qquad r_2(0) = I, \qquad t \in R^+ \tag{1.4}$$

respectively. Here, $r_j(t)$ ($j = 1, 2$), A and $B(t)$ are n by n matrices, I is the identity matrix, and $x(t)$ and $f(t)$ are n-dimensional row vectors. The

**Current affiliation*: Department of Mathematics, Virginia Polytechnic Institute and State University, Blacksburg, Virginia

importance of these resolvent kernels stems from the fact that under mild assumptions equations (1.1) and (1.2) are respectively solved by

$$x(t) = f(t) - \int_0^t f(t - s)r_1(s)ds, \qquad t \in R^+ \tag{1.5}$$

$$x(t) = x_0 r_2(t) + \int_0^t f(t - s)r_2(s)ds, \qquad t \in R^+ \tag{1.6}$$

In addition, $r_1(t)$ and $r_2(t)$ occur in "variation of constants" formulae (see [10, Chapter 4] and [4]) associated with certain nonlinear perturbed forms of the systems (1.1) and (1.2).

The purpose of this paper, which summarizes recent joint work with G. S. Jordan [8], is to give conditions that ensure that for j = 1, 2 the resolvent kernel $r_j(t)$ has the form

$$r_j(t) = \sum_{k=1}^{M} P_k(t)e^{z_k t} + u_j(t), \qquad t \in R^+ \tag{1.7-j}$$

where, for each k, $P_k(t)$ is a matrix of polynomials, and $u_j(t)$ belongs to a weighted space $L^1(R^+;\rho)$. If the resolvent kernel $r_1(t)$ (resp. $r_2(t)$) is known to have the form (1.7-1) (resp. (1.7-2)), then the representation (1.5) (resp. (1.6)) can be used to draw a number of conclusions concerning the asymptotic behavior of the solution x(t) of (1.1) (resp. (1.2)) from the corresponding behavior of f(t); see, e.g., [12, Section 6]. For proofs and an expanded discussion of our results we refer the reader to [8].

By a weight function $\rho(t)$ we mean a positive continuous function on R^+ such that $\rho(0) = 1$,

$$\rho(s + t) \le \rho(s)\rho(t), \qquad 0 \le s, t < \infty \tag{1.8}$$

and, if

$$\rho_* \equiv \lim_{t\to\infty} \frac{\log \rho(t)}{-t} \qquad \text{with} \qquad -\infty < \rho_* < \infty$$

then

$$\rho(t)e^{\rho_* t} \quad \text{is nondecreasing on } R^+ \tag{1.9}$$

(The existence of the limit ρ_* follows from (1.8); see [2, p. 113]. The regularity condition (1.9) is imposed in order to estimate certain integrals that occur in our proofs.) An interesting example of a weight function $\rho(t)$ that satisfies (1.8) and (1.9) is given by

$$\rho(t) \equiv e^{-\rho_* t}(1 + t)^{\delta}[1 + \log(1 + t)]^{\gamma}, \qquad t \in R^+$$

with $-\infty < \rho_* < \infty$ and $\delta, \gamma \geq 0$.

The space $L^1(R^+;\rho)$ consists of all n by n matrix functions B(t) such that each component B_{ij} is measurable and satisfies

$$\int_0^{\infty} \rho(t)\ |B_{ij}(t)|dt < \infty \qquad 1 \leq i, \qquad j \leq n$$

If $B(t) \in L^1(R^+;\rho)$, the Laplace transform

$$\hat{B}(z) \equiv \int_0^{\infty} e^{-zt}\ B(t)dt$$

exists and is continuous on Re $z \geq \rho_*$ and analytic in Re $z > \rho_*$. Hence, the meaning of a zero of order m of $\det[I + \hat{B}(z)]$ (resp. $\det[zI - A - \hat{B}(z)]$) is clear when Re $z > \rho_*$. If z_0 on Re $z = \rho_*$ is a zero of one of these determinants, we say that z_0 is a zero of order m if the first m - 1 derivatives of the determinant are zero, but the mth derivative is nonzero. These m derivatives exist when, say, $(1 + t)^m\ B(t) \in L^1(R^+;\rho)$.

THEOREM 1. Let m be a nonnegative integer and assume that $(1 + t)^{2m}\ B(t) \in L^1(R^+;\rho)$. Assume that the only zeros of $\det[I + \hat{B}(z)]$ in Re $z \geq \rho_*$ occur at $z = z_k$, $1 \leq k \leq M < \infty$. Let m_k be the order of the zero at z_k, and if Re $z_k = \rho_*$, assume that $m_k \leq m$. Then the solution $r_1(t)$ of equation (1.3) has the form (1.7-1) where, for each k, $P_k(t)$ is a matrix of polynomials of degree at most $m_k - 1$ which depends only on B(t), and $u_1(t) \in L^1(R^+;\rho)$.

The corresponding result for the differential resolvent $r_2(t)$ is

THEOREM 2. Let the hypotheses of Theorem 1 hold with the exception that $\det[I + \hat{B}(z)]$ in the statement of Theorem 1 is now replaced by $\det[zI - A - \hat{B}(z)]$. Then the solution $r_2(t)$ of equation (1.4) has the form (1.7-2) with both $u_2(t)$ and $u_2'(t)$ in $L^1(R^+;\rho)$.

In the case where $\rho(t) \equiv 1$, that is when $L^1(R^+;\rho)$ is the usual Lebesgue space $L^1(R^+)$, the scalar case of Theorem 2 was obtained by Miller [12, Theorem 6]. If equation (4) is not scalar, then Theorem 2 (with $\rho(t) \equiv 1$) sharpens Theorem 5 of [12] in which the kernel B(t) is required to have $2M_1$ moments where M_1 is the sum of the multiplicities of the zeros of $\det[zI - A - \hat{B}(z)]$ which lie on Re $z = \rho_*$.

Recently, Hannsgen has obtained a very general Wiener-Lévy Theorem for quotients [6, Theorem 1.1] which yields as easy corollaries generalizations of Theorems 1 and 2 in the case where the weight function $\rho(t) \equiv 1$. In these results of Hannsgen, the moment hypothesis on B(t) may be replaced by a combination of moment and monotonicity and convexity assumptions. A particular case of the results in [6] for the scalar equation (1.4) with $A = 0$ and with B(t) belonging to a special class of piecewise linear kernels was obtained by Hannsgen in [5].

Finally, we remark that the special cases of Theorems 1 and 2 in which the appropriate determinant has no zeros in Re $z \geq \rho_*$ are due to Gelfand, Raikov and Shilov [2, p. 116] (see also [1]) and Shea and Wainger [13], respectively. For other recent results that imply that $r_j(t)$ belongs to a weighted space $L^1(R^+;\rho)$, see Gripenberg [3] and Jordan and Wheeler [9].

Theorems 1 and 2 are easy consequences of results for more general linear Volterra-Stieltjes systems of convolution type which we now discuss.

2. VOLTERRA-STIELTJES SYSTEMS

Consider the linear Volterra-Stieltjes systems

$$\int_0^t x(t - s)dA(s) = f(t), \qquad t \in R^+ \tag{2.1}$$

$$x'(t) + \int_0^t x(t - s)dA(s) = f(t), \qquad x(0) = x_0, \qquad t \in R^+ \tag{2.2}$$

where $f = (f_1, \ldots, f_n)$ and x are complex vector functions with n components, and $A = [A_{ij}]$ is an n by n matrix of complex-valued functions.

Let $\rho(t)$ be a weight function as in Section 1. The weighted space $V_+[\rho]$ consists of all n by n matrix functions A(t) such that each component A_{ij} is of bounded variation on every finite interval [0,T], is normalized to be left-continuous and vanish at 0, and satisfies

$$\|A_{ij}\| \equiv \int_0^\infty \rho(t)|dA_{ij}(t)| < \infty \qquad 1 \le i, j \le n$$

For $A \in V_+[\rho]$, the Laplace-Stieltjes transform $\tilde{A}(z) \equiv \int_0^\infty e^{-zt}dA(t)$ is continuous on Re $z \ge \rho_*$ and analytic in Re $z > \rho_*$.

Associate with $A(t) \in V_+[\rho]$ the scalar function $D(t) \in V_+[\rho]$ which satisfies

$$\tilde{D}(z) = \det \tilde{A}(z), \text{ Re } z \ge \rho_* \tag{2.3}$$

As in Section 1, we say that z_0 on Re $z = \rho_*$ is a zero of $\tilde{D}(z)$ of order m if

$$\tilde{D}(z_0) = \tilde{D}'(z_0) = \dots = \tilde{D}^{(m-1)}(z_0) = 0, \qquad \text{but } \tilde{D}^{(m)}(z_0) \ne 0$$

Recall (see [1] or [2, p. 166]) that a function $A(t) \in V_+[\rho]$ may be decomposed uniquely as

$$A(t) = h_A(t) + g_A(t) + s_A(t) \tag{2.4}$$

where $h_A = [h_{A_{ij}}]$ is a matrix of discrete functions, g_A is a matrix of functions absolutely continuous on each finite interval, and s_A is a matrix of singular functions.

With this background we are ready to state our main result which concerns equation (2.1)

THEOREM 3. Let m be a nonnegative integer, $A(t) \in V_+[\rho]$ and $D(t)$ be defined as in (2.3). Assume that the only zeros of $\tilde{D}(z)$ in Re $z \ge \rho_*$ occur at $z = z_k$, $1 \le k \le M < \infty$. Let m_k be the order of the zero at z_k and if Re $z_k = \rho_*$, assume that $m_k \le m$. Assume that $A(t)$ and $f(t)$ satisfy the respective moment hypotheses

$$\int_0^\infty \rho(t)t^{2m}|dA_{ij}(t)| < \infty \qquad 1 \le i, j \le n \tag{2.5}$$

$$\int_0^\infty \rho(t)(1 + t)^m|f_i(t)|dt < \infty \qquad 1 \le i \le n \tag{2.6}$$

Finally, assume that

$$\left\{\begin{array}{l}\text{except near points } z_k,\ 1 \le k \le M,\\ 1/\tilde{D}(z) \text{ is bounded in } \operatorname{Re} z \ge \rho_*\end{array}\right. \tag{2.7}$$

and

$$\inf_{-\infty<\sigma<\infty} |\det \tilde{h}_A(\rho_* + i\sigma)| > \| s_D \| \tag{2.8}$$

If x(t) is a Borel measurable solution of (2.1) which is integrable on each finite interval, then x(t) has the form

$$x(t) = \sum_{k=1}^{M} p_k(t)e^{z_k t} + x_1(t) \tag{2.9}$$

where, for each k, $p_k(t)$ is a vector whose entries are polynomials of degree at most $m_k - 1$ which depends on A and f, and $x_1(t) \in L^1(R^+;\rho)$.

The corresponding result for equation (2.2) may be deduced from Theorem 3 using a technique introduced by Miller [11].

THEOREM 4. Let m be a nonnegative integer, $A(t) \in V_+[\rho]$, and assume that A(t) and f(t) satisfy (2.5) and (2.6), respectively. Assume that the only zeros of $\det[zI + \tilde{A}(z)]$ in $\operatorname{Re} z \ge \rho_*$ occur at $z = z_k$, $1 \le k \le M < \infty$, let m_k be the order of the zero at z_k, and if $\operatorname{Re} z_k = \rho_*$, assume that $m_k \le m$. If x(t) is a locally absolutely continuous function such that $x(0) = x_0$ and (2.2) holds a. e. on R^+, then x(t) has the form (2.9) with $x_1(t)$ and $x_1'(t)$ both in $L^1(R^+;\rho)$.

Observe that the hypotheses of Theorem 4 do not include a restriction such as (2.8) on the singular part of A(t).

Theorems 3 and 4 are similar to results obtained by Jordan and Wheeler in [7] where it is proved, under different moment hypotheses, that if f(t) is bounded on finite intervals and $f(t) = o(1/\rho(t))$ as $t \to \infty$, then the solutions x(t) of (2.1) and (2.2) have the form (2.9) with $x_1(t) = o(1/\rho(t))$ as $t \to \infty$. However, Theorems 3 and 4 are more useful than the results in [7]. In particular, Theorems 1 and 2 concerning the resolvents

$r_j(t)$ $(j = 1, 2)$ follow from Theorems 3 and 4, respectively, and the conclusion that $r_j(t)$ $L^1(R^+;\rho)$ has wider applicability than $r_j(t) = o(1/\rho(t))$ as $t \to \infty$.

ACKNOWLEDGMENT

Partially supported by National Science Foundation Grant MCS78-01330.

REFERENCES

1. I. M. Gelfand, Über absolut konvergente trigonometrische Reihen und Integrale, *Mat. Sb.* 9 (1941), 51-66.

2. I. M. Gelfand, D. A. Raikov, and G. E. Shilov, *Cummutative Normed Rings*, Chelsea, New York, 1964.

3. G. Gripenberg, On rapidly decaying resolvents of Volterra equations, *J. Integral Equations*, 1 (1979), 241-247.

4. S. I. Grossman and R. K. Miller, Perturbation theory for Volterra integrodifferential systems, *J. Differential Equations*, 8 (1970), 457-474.

5. K. B. Hannsgen, An L^1 remainder theorem for an integrodifferential equation with asymptotically periodic solution, *Proc. Amer. Math. Soc.*, 73 (1979), 331-337.

6. K. B. Hannsgen, A Wiener-Lévy Theorem for quotients, with applications to Volterra equations, *Indiana Univ. Math. J.*, 29 (1980), 103-120.

7. G. S. Jordan and R. L. Wheeler, Asymptotic behavior of unbounded solutions of linear Volterra integral equations, *J. Math. Anal. Appl.*, 55 (1976), 596-615.

8. G. S. Jordan and R. L. Wheeler, Structure of resolvents of Volterra integral and integrodifferential systems, *SIAM J. Math. Anal.*, 11 (1980), 119-132.

9. G. S. Jordan and R. L. Wheeler, Rates of decay of resolvents of Volterra equations with certain nonintegrable kernels, *J. Integral Equations*, 2 (1980), 103-110.

10. R. K. Miller, *Nonlinear Volterra Integral Equations*, W. A. Benjamin, Menlo Park, Ca., 1971.

11. R. K. Miller, Asymptotic stability and perturbations for linear Volterra integrodifferential systems, *Delay and Functional Differential Equations and Their Applications*, Academic Press, New York, 1972.

12. R. K. Miller, Structure of solutions of unstable linear Volterra integrodifferential equations, *J. Differential Equations*, 15 (1974), 129-157.

13. D. F. Shea and S. Wainger, Variants of the Wiener-Lévy Theorem, with applications to stability problems for some Volterra integral equations, *Amer. J. Math.*, 97 (1975), 312-343.